PLANT AUTOGRAPHS AND THEIR REVELATIONS

PLANT AUTOGRAPHS AND THEIR REVELATIONS

JAGADIS CHUNDER BOSE

GYAN

Published by

GYAN

Gyan Publishing House
5, Ansari Road
Daryaganj, New Delhi-110002
Phone: 011-47034999, 9811692060
E-mail: books@gyanbooks.com

Distribution Network
gyanbooks.com
India, USA, Canada, UK, Australia

ISBN: 978-93-7414-294-3 (PB)

First Published, 1927

2nd Impression 2026

Printed at: Gyan Press, Delhi.

PLANT AUTOGRAPHS AND THEIR REVELATIONS
Author: JAGADIS CHUNDER BOSE

SIR J. C. BOSE, M.A., F.R.S.

PLANT AUTOGRAPHS AND THEIR REVELATIONS

BY

SIR JAGADIS CHUNDER BOSE

M.A., D.SC., LL.D., F.R.S., C.S.I., C.I.E.
MEMBER OF INTERNATIONAL COMMITTEE
ON INTELLECTUAL CO-OPERATION,
LEAGUE OF NATIONS,
FOUNDER AND DIRECTOR BOSE INSTITUTE,
CALCUTTA

WITH PORTRAIT AND 120 ILLUSTRATIONS IN THE TEXT

LONGMANS, GREEN AND CO. LTD.
39 PATERNOSTER ROW, LONDON, E.C. 4
NEW YORK, TORONTO
CALCUTTA, BOMBAY AND MADRAS
1927

To

MY WIFE

WHO HAS STOOD BY ME IN ALL
MY STRUGGLES

PREFACE

THE present volume gives a connected and popular account of the researches that have been carried out in the Bose Institute, one of the objects of which, as explained in the Appendix, is to associate the advancement of knowledge with the widest possible civic and public diffusion of it.

In pursuing investigations on the border region of physics and physiology, I was amazed to find boundary lines vanishing and points of contact emerge between the realms of the Living and Non-Living. Metals are found to respond to stimuli; they are subject to fatigue, stimulated by certain drugs and 'killed' by poisons.

Between inorganic matter at one extreme end and animal life at the other, there is spread out the vast expanse of the silent life of plants. The difficulty that thwarts the investigator at every step arises from the fact that the interplay of life-action is taking place within the dark profundities of the tree, which our eyes cannot penetrate. In order to reveal the intricate mechanism of its life, it is necessary to gain access to the smallest unit of life, the 'life-atom,' and record its throbbing pulsation. When microscopic vision fails we have still to explore the realm of the invisible.

I have been able to make the dumb plant the most eloquent chronicler of its inner life and experiences by making it write down its own history. The self-made records thus made show that there is no life-reaction in even the highest animal which has not been foreshadowed in the life of the plant.

I shall take my readers with me step by step as the

wonders became gradually revealed to me through artificial organs of extraordinary sensitiveness by which alone the realm of the invisible could be explored. The barriers which seemed to separate kindred phenomena will be found to have vanished, the plant and the animal appearing as a multiform unity in a single ocean of being. In this vision of truth the final mystery of things will by no means be lessened, but greatly deepened. It is not less of a miracle that man, circumscribed on all sides by the imperfections of his senses, should yet build himself a raft of thought to make daring adventures in uncharted seas. And in his voyage of discovery he catches an occasional glimpse of the ineffable wonder that had been hidden from his view. That vision crushes out of him all self-sufficiency, all that kept him unconscious of the great pulse that beats through the universe. The joy that fills my life I would wish to share with my readers.

J. C. Bose.

Bose Institute, Calcutta,
October, 1926.

CONTENTS

CHAPTER I

THE SILENT LIFE

CHAPTER II

PLANT-SCRIPT

CHAPTER III

THE BEHAVIOUR OF PLANTS

CHAPTER IV

DRUGGED PLANTS

CHAPTER V

THE ELECTRICAL RESPONSE

CHAPTER VI

THE SLEEP OF PLANTS

CHAPTER VII

THE PRAYING PALM OF FARIDPORE

CHAPTER VIII

RESPONSE OF INORGANIC MATTER

CHAPTER IX

THE CURVE OF LIFE AND DEATH

CHAPTER X

AUTOMATISM

CHAPTER XI

THE RECORD OF GROWTH

CHAPTER XII

THE MAGNETIC CRESCOGRAPH

CHAPTER XIII

WOUNDED PLANTS

CHAPTER XIV

THE SENSE OF DIRECTION IN PLANTS

CHAPTER XV

PLANT MOVEMENTS IN NATURE

CHAPTER XVI

THE NIGHT-WATCH OF NYMPHÆA

CHAPTER XVII

THE RISE OF THE SAP

CHAPTER XVIII

THE PROPULSIVE TISSUE

CHAPTER XIX

THE WEEPING MANGO-TREE

CHAPTER XX

MILKING THE PALM-TREE

CHAPTER XXI

ACTION OF ALKALOIDS AND COBRA-VENOM ON PULSE-BEAT OF ANIMAL AND PLANT

CHAPTER XXII

CARBON-ASSIMILATION

CHAPTER XXIII

THE NERVE OF PLANTS

CHAPTER XXIV

LOCALISATION OF THE NERVE

CHAPTER XXV

HELIOS AND HIS LEAF-CHARIOT

CHAPTER XXVI

DISCOVERY OF A REFLEX ARC IN PLANTS

CHAPTER XXVII

THE CONTROL OF NERVOUS IMPULSE

APPENDIX

THE BOSE INSTITUTE

PLANT AUTOGRAPHS AND THEIR REVELATIONS

CHAPTER I

THE SILENT LIFE

THE immense possibilities of life in the plant are at times latent and all but imperceptible. The gigantic Banyan tree exists potentially in a body smaller than a mustard seed. Within the enclosure of the seed-coat life lies long in a state of suspense, to be awakened at the proper season and under favourable conditions. Extraordinary is the power of resistance possessed by life in its latent phase. Seeds have been subjected to cold so intense that under its action air becomes liquefied and mercury solid; and yet, within the defensive armour of the seed-coat, life persists, resisting all dangers from outside. Storms and hurricanes scatter the living seeds. The coco-nut rides the waves till anchored safely in some island yet to be inhabited. In due season there commences a series of astonishing transformations; the latent life awakens, the seed-coat bursts asunder and growth begins. The seedling seems as if possessed of faculties almost human. Place it upside down, yet the primary root bends down unerringly to pierce the earth for water, and the stem rises upward for light and air. The root winds its way underground, avoiding stones and obstacles. The stem bends towards the light, and tendrils twine around a support. Such visible movements are striking enough. But within the unruffled exterior of the plant-body, there are other activities, energetic and incessant, which have hitherto escaped scrutiny. Until we can enter into this realm of the invisible, the mystery of its life and its diverse manifestations will remain unsolved.

The plant has always appeared to us very remote, because its life is so still and silent. In its apparent immobility and placidity the plant stands in strong contrast to the energetic animal with its reflex movements and pulsating organs. Yet the same environment which, with its changing influences, so profoundly affects the animal, is also playing upon the plant. Storm and sunshine, the warmth of summer and the frost of winter, drought and rain—all these and many more come and go about it. What subtle impressions do they leave behind? Internal changes there must be; but our eyes have not the power to see them.

The life-activities of plants appear, at first sight, to be very different from those of animals. But if, in spite of the seeming differences, it could be proved that they are fundamentally similar, that would undoubtedly constitute a scientific generalisation of very great importance. It would then follow that the complex mechanism of the animal machine that has baffled us so long, need not remain inscrutable for all time, since the intricate problems of animal life would naturally find their solution in the study of corresponding problems in the simpler vegetable life. It would mean the possibility of very great advance in the science of physiology, of agriculture, of medicine, and even of psychology.

We are still profoundly ignorant of the fundamental reactions which underlie the life of plants, for numerous are the difficulties which baffle the investigator. Such knowledge as we have consists of insufficient facts upon which contending and irreconcilable theories have been based. Under these circumstances the only clear path is that which leads us away from the din of disputation to follow the trail of fact. We must therefore abandon all our preconceptions and put our questions directly, insisting that the only evidence which can be accepted is that which bears the plant's own signature.

There are professors of a science bordering on the mystical who declare that they can tell the character and disposition of any one simply by careful observation of his handwriting. Scepticism is permissible as to the authenticity of such claims; yet there is no doubt that one's handwriting may be modified profoundly by conditions, physical and mental. There still exist at Hatfield House documents signed by Guy Fawkes. Those who have seen them declare that there is a sinister variation in these signatures. The crabbed and distorted characters of the last words Guy Fawkes wrote on earth as, in the dark hours of the morning on which he was executed, he set his hand to the written confession of his crime, tell their own tale of what had transpired in the solitary hours of that fateful night.

If such is the revelation that may be unfolded to the critical eye by the lines and curves of a human autograph, perhaps the plant can be induced to give us its own autograph which would similarly reveal its internal condition. The only conceivable way of effecting this would be to obtain a record of the response of the organism to a questioning shock. When an animal receives an external shock, it may answer in various ways; if it has a voice, by a cry; if dumb, by the movement of its limbs. The external shock is the stimulus; the answer of the organism is the response. If we can make the plant likewise give some perceptible answer in response to a shock, we can then judge of the condition of the plant by the character of the answer. When in an excitable condition, the feeblest stimulus will evoke an extraordinarily strong response; in a depressed state even a strong stimulus evokes only a feeble reply; and lastly, when death overcomes life, there is an abrupt end of the power to answer at all.

So long as we are alive we respond in one way or another to the various shocks or stimuli of our environment. Stimulus falling on matter produces a molecular upset, causing excitation, and this upset is exhibited in various ways according to the instrument of expression. Let us look for a moment at the different effects which may be produced by the same electrical current acting on different instruments. Acting on one kind of recorder, it produces a movement; on another, say an electric bell, it produces sound; on still another, it causes a flash of light. In like fashion, a stimulus acting on living tissue may cause excitation which manifests itself in different ways, according to the different organs of expression.

Take first the familiar example of response by movement; a drop of boiling water falling on the hand causes the muscles to twitch and the hand to be withdrawn. The same thing is seen in the leaf of the sensitive plant Mimosa, which suddenly falls down when struck. Also under the excitation of touch, the open Dionaea leaf closes on the fly, its prey. Again, instead of this mechanical movement we may, as we shall see, have electrical movement in response to stimulus; and a suitable electrical recorder, the galvanometer, shows an electrical twitch each time the living tissue is excited. And lastly, among the forms of response, we have sensation itself.

Sight is the characteristic response of the retina; hearing of the ear, and so on; and in sensation, different parts of the brain act as so many responding organs. For the purpose of investigation in the laboratory we may take, say, the eye of a frog, with its optical nerve connected with the galvanometer instead of with the brain. It will then be found that each time a flash of light falls on the retina, the galvanometer responds with a sudden deflection, just as specific reactions in the brain are perceived as sensations.

We thus see the possibility of obtaining some kind of record of the response of the plant, by which its internal condition and the changes of that condition may be revealed. In order to succeed in this, we have first to discover some compelling force that will make the plant give an answering signal, mechanical or electrical; secondly, to supply a means for the conversion of these signals into an intelligible script; and last of all, we have ourselves to learn the specific significance of this hieroglyphic.

In answering the question, then, as to whether there is an essential unity in the responsivity of plant and animal, we have first to find out whether sensitiveness is characteristic of only a few plants, or whether all plants and every organ of every plant are sensitive. We have then to compare the characteristic reactions of plant and animal under parallel conditions, in order to find whether the normal response and its modifications under changed conditions are similar in the two cases.

When an animal receives a blow, it does not respond immediately. A certain short interval elapses between the incidence of the blow and the beginning of the reply. This lost time is known as the *latent period*. The psycho-analyst learns more of his human patient from this pause than his patient can tell him in words. In the plant also a definite period elapses between the incident blow and the responsive twitch. Does this latent period undergo variation, as in the animal, with change in the internal or external conditions? Is it possible to make the plant itself write down for us this very short latent period?

Further, is the plant excited by the various irritants which excite the animal? Is this irritation local, or does it give rise to an impulse which travels through the length of the plant and causes movement at a distance? If so, at what rate does this impulse travel, and what are the conditions which increase the rate, and under what other conditions is the impulse retarded or arrested? Is it possible to make the plant itself record the rate and its variations? Is there any resemblance between the nervous impulse in the animal and the excitatory impulse in the plant?

The characteristic effects of various drugs are more or less well known in the case of the animal. Is the plant similarly susceptible to their action? Will the effect of poison change with the dose? Is it possible to counteract the effect of one poison by means of another which acts as an antidote?

In the animal there are certain pulsating tissues such as that of the heart. Are there any such spontaneously beating tissues in the plant? If so, are the automatic pulsations in the animal and in the plant affected by external conditions in a similar manner? And what is the real meaning of spontaneity or automatism?

Growth furnishes us with the most important example of automatism. But the rate of growth in a plant is far below anything that we can directly perceive. How then is this growth to be magnified so as to be rendered instantly measurable? What are the variations in this infinitesimal movement under external stimulus of light, or of touch, or of the shock of the electric current? What changes are produced by giving or withholding water and by application of various drugs? Above all, what are the main conditions which stimulate or retard growth?

And lastly, when, in any of the manifold ways, life is finally extinguished, will it be possible to detect the critical moment? Does the dying plant give any unmistakable sign at the moment after which its activities cease for ever?

CHAPTER II

THE PLANT-SCRIPT

WITH all the preceding questions before us, we have to face the difficulty that plants are dumb and apparently passive creatures; yet may they not give us some signals from which we can read their inner history? Dumb people make themselves understood by movements of their fingers; let us see how we can learn something of their vital condition. We tap a dumb person on the finger; that will be the questioning shock or stimulus, and the finger twitches in answer. But to what extent? That will depend on how much alive he is. If he be very much alive, the twitch of his finger will be strong; if depressed, the same blow will cause a feeble twitch; if he is dead, there will be no answering twitch at all.

Suppose we had tied a pencil to his finger and had a piece of paper moving past the pencil. We should, in the experiments described, get three kinds of record. If the dumb person were fully alive, the line recorded by the twitch would be long; if in a depressed condition, the answering record would be short, and in death there would be no recorded movement. Thus from the length of the lines we should be able to judge of the condition of this dumb person.

Somewhat in the same way we shall be able to discover the vitality of the plant, by the amount of movement made in answer to a testing shock. While all plants have a limited power of movement, none have such active movements, comparable to those of animals, as the so-called 'sensitive' plants in general and *Mimosa pudica* in particular. A representation of a pair of leaves of Mimosa is given in fig. 1: the left leaf is in the normal outspread position; the right leaf has fallen down under a shock.

It is a very fascinating sight to watch the movements of the Mimosa. In Bengal it is called the 'coy maiden,' because of its sensitive shrinking away from all touch. Children love to play with it by the hour, attracted by the impulsive movement in answer to every kind of irritation.

The entire leaf of the Mimosa may be likened to our own horizontally outstretched arm with the fingers outspread. In order to avoid the use of long descriptive phrases, it is advisable to use short names for different parts of the leaf. The main leaf-stalk or *petiole* is like the arm,

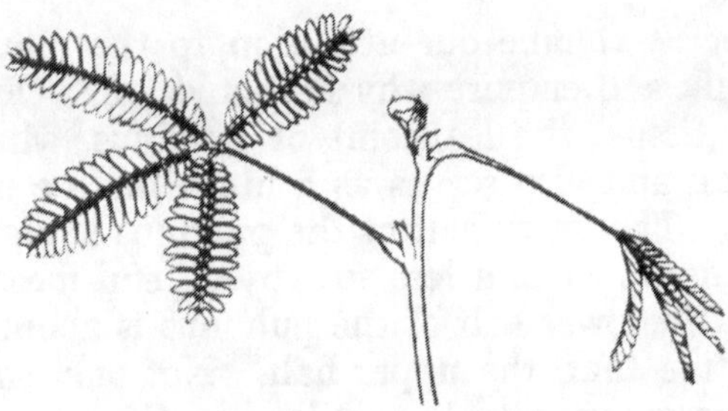

FIG. 1. Leaves of Mimosa in expanded condition (left) and contracted condition after stimulation (right).

and the four *sub-petioles* are like the outspread fingers. Strike the arm, and it bends in contraction and the fingers close; so also a Mimosa leaf, when struck, falls down, and the sub-petioles are drawn close together. The sub-petioles bear numerous pairs of leaflets which are also sensitive and close upwards. The movement is brought about by contraction of the sensitive cushion-like mass of tissue, the *pulvinus*, at the different joints, which contracts like the muscle of the animal. There is a large pulvinus at the junction of the leaf with the stem; there are four smaller pulvini at the junctions of the four sub-petioles with the main petiole, and there are numerous minute pulvinules at the junction of the leaflets with the sub-petioles.

It is a wonderful leaf-mechanism, that of the Mimosa;

three distinct and differing motions in one single leaf. The main leaf-stalk drops down; the four sub-petioles are moved laterally and drawn together; and the small pairs of leaflets fold upwards. Is it any wonder that we should be fascinated by such a lively display of movement in a plant?

If any one of the terminal leaflets be pinched or cut with a pair of scissors or touched with a hot wire, an impulse is found to travel inwards, that is to say, towards the stem. The different pairs of leaflets fold themselves upward one after another in regular succession; the sub-petioles are drawn together, and the main petiole drops down. If the irritation has been very great, the excitation passes along the stem, causing the fall of other distant leaves.

Now let us confine our attention to the movement of the leaf-stalk and enquire why it should fall. Observe the cushion of tissue, the leaf-joint or pulvinus, which is the motile organ and also serves as a hinge for the movement of the leaf. The lower half of the pulvinus is highly sensitive, and the upper half less so: by careful measurement, I find that the lower half of the pulvinus is about 80 times more sensitive than the upper half. Not only so, but the lower half is more actively motile than the upper. Hence when we stimulate the leaf, both halves contract, but the greater contraction of the lower half causes the leaf to fall. The actual shortening of the lower half of the pulvinus is very small, but as the long leaf-stalk acts as a magnifying index, the movement of response becomes very conspicuous.

The Response Recorder

I shall now describe the method of recording the responsive movement under stimulation. A leaf of a Mimosa is attached by means of a fine thread to one end of the lever (fig. 2), which is pivoted on jewel bearings. The other end of the lever carries a small weight by which the string is kept taut. From the middle of the lever and at right angles to it, there extends a thin wire with a curved end serving as the writer. The tip of this wire just touches a smoked glass plate on which the record is made, which is allowed to slide down under the action of gravity, at a rate regulated by clockwork.

We may stimulate the plant by employing agents that excite human beings—a pinch, the application of a hot wire or of a drop of acid. These, however, may readily cause serious injury to the plant, and are, therefore, unsuitable, especially for repetition. A better way is to employ an electrical shock from an induction coil, which has the following advantages: first, it does not injure the plant, provided the shock is not too intense; second, successive shocks can be kept absolutely constant; and third, the intensity of the shock can, when necessary, be gradually varied. In the induction apparatus there are two coils, a primary and a secondary. The intensity of the shock can be gradually increased by bringing the secondary nearer to the primary coil (fig. 3).

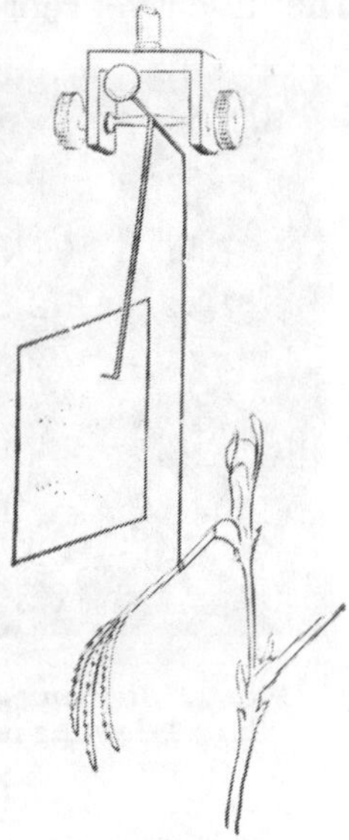

FIG. 2. Diagrammatic representation of Response Recorder.

When the leaf falls on excitation it pulls down the right end of the lever, so that the writer

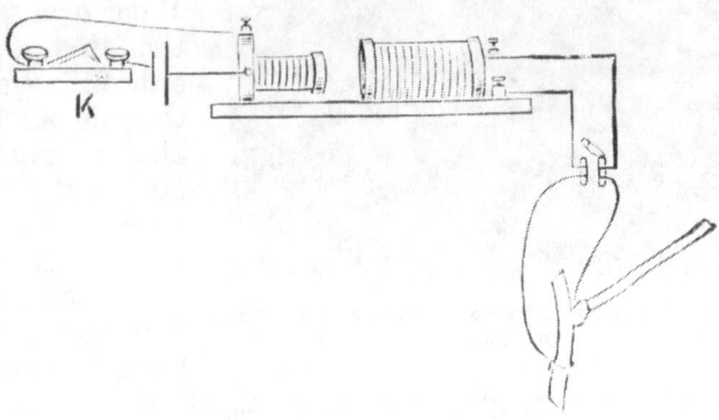

FIG. 3. Arrangement for applying induction shock. K, key in the primary circuit.

moves to the left, marking an up-line. The leaf slowly recovers from the effect of the shock and again becomes erect, producing the down-curve of recovery. The up- and down-curves together constitute a single pulse of response. The up-curve, representing contraction, is completed in a

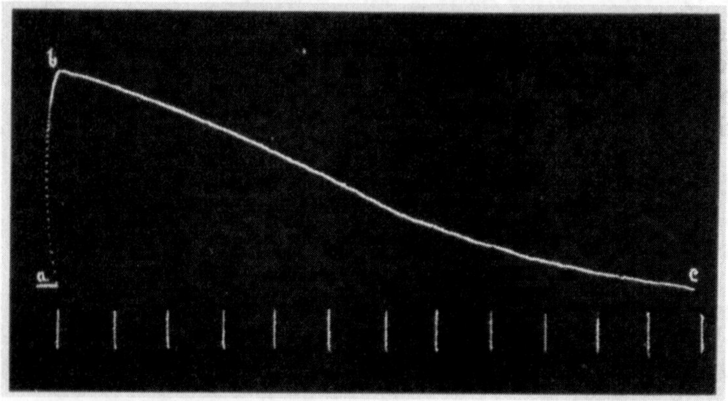

FIG. 4. Response-curve of primary leaf of Mimosa. The vertical lines below the record indicate intervals of one minute each. (*a b*), fall of leaf; (*b c*), recovery.

little over a second, whereas the recovery takes as long as twelve minutes.

In fig. 4 is shown a record of the response of Mimosa; it also represents vividly the course of our own sensation after getting a shock. We feel the pain of a blow almost immediately, but it takes some time for the painful sensation to disappear. After the Mimosa has completed its recovery, we may give it a second shock of the same intensity. If the surrounding conditions are the same as before, the second response is the same, and so are the succeeding ones (fig. 5). Should the plant be depressed in any way, the fact is at once revealed by depression in its responses.

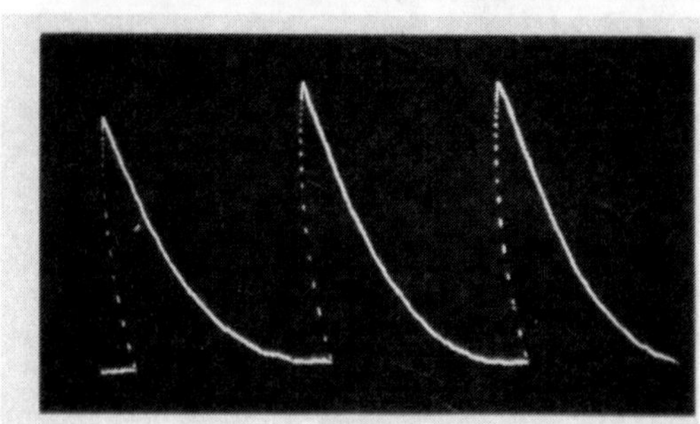

FIG. 5. Record of successive uniform responses of Mimosa.

The Resonant Recorder

I have not yet referred to a serious difficulty in obtaining accurate plant-records. In the case of muscle, the force of contraction is considerable, so the friction offered by the recording surface presents no serious difficulty. In the case of plants, however, the pull exerted by the motile

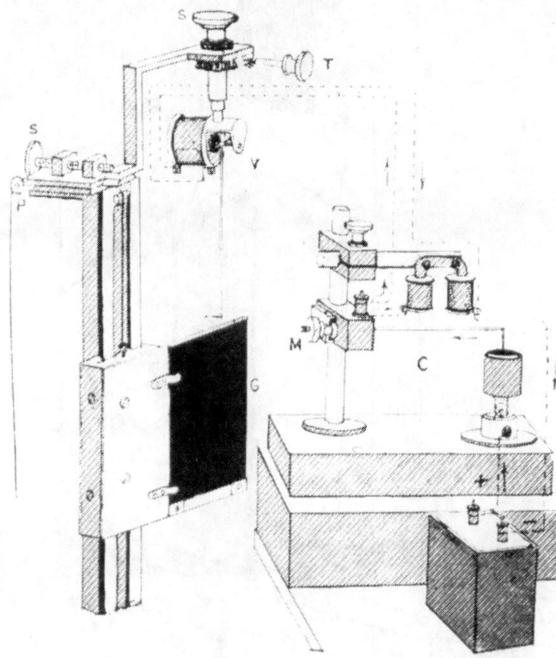

Fig. 6. Upper part of Resonant Recorder. (From a photograph.)

organ is feeble. If the lever were allowed to remain in con tinuous contact with the smoked glass, the pressure would be enough to modify or arrest the movement. This difficulty is overcome by making an intermittent instead of a permanent contact, thus giving rise to a series of dots in the record instead of a continuous line. This has been accomplished by my Resonant Recorder (fig. 6), the principle of which is based on sympathetic vibration.

If the strings of two violins are exactly tuned, then a note sounded on one will cause the other to vibrate in

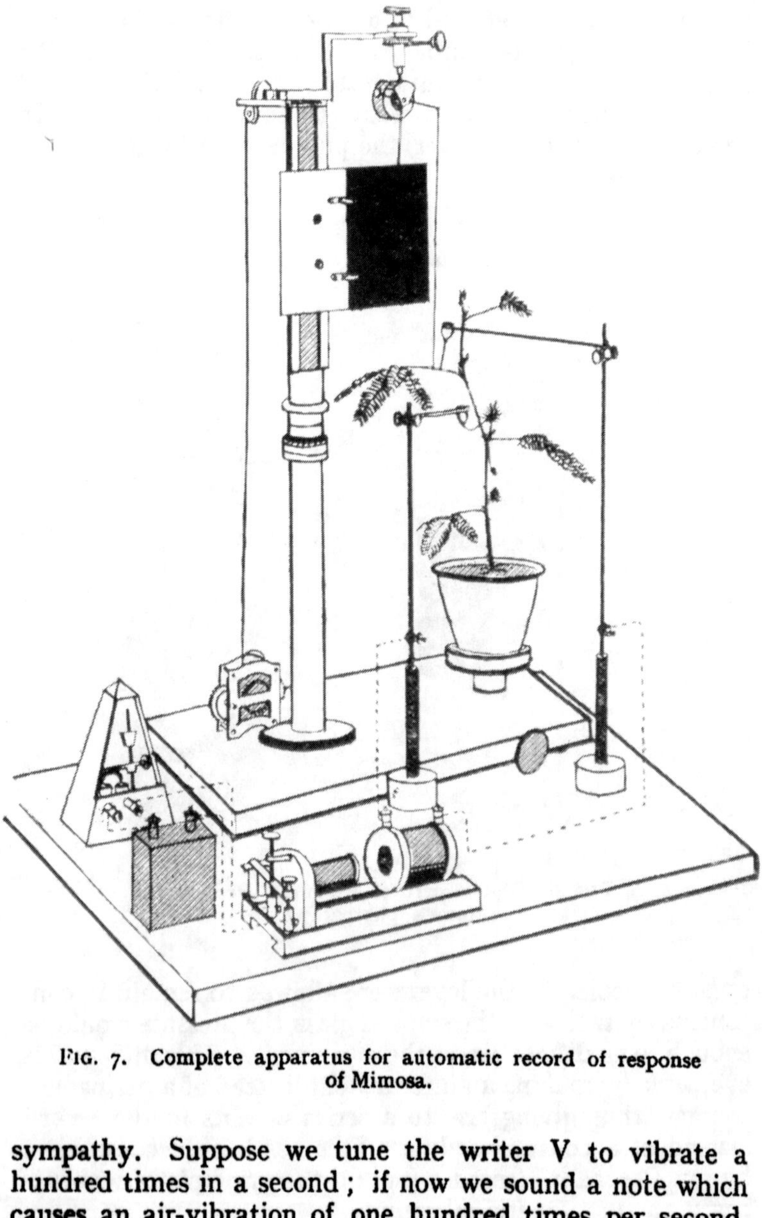

FIG. 7. Complete apparatus for automatic record of response of Mimosa.

sympathy. Suppose we tune the writer V to vibrate a hundred times in a second; if now we sound a note which causes an air-vibration of one hundred times per second, the writer will vibrate in sympathy. It will no longer remain in continuous contact with the recording plate, but will deliver a succession of taps a hundred times a second. The record will, therefore, consist of a series of dots, the distance between one dot and the next representing an interval of one hundredth part of a second.

With finer recorders it is possible to measure intervals of time as short as a thousandth part of a second. It will now be understood how, by the device of the Resonant Recorder, not only is the difficulty due to friction removed, but the record itself is made to measure the time, however short the intervals required may be..

In this way I have succeeded in making the plant itself write an answering script to a definite testing stimulus; and in order that the results obtained should not be influenced by any personal bias, arrangements were made that the plant attached to the recording apparatus should be automatically excited by an absolutely constant stimulus, and should make its own responsive record, going through its own period of recovery and repeating the same cycle without assistance or interference at any point on the part of the observer. Fig. 7 illustrates the complete apparatus.

Sensitive and Ordinary Plants

We have been accustomed to regard the plant-world as insensitive, Mimosa and a few other 'sensitive plants' being considered to be exceptions. Our justification for dividing plants into these two classes is that while Mimosa makes an active movement when struck, ordinary plants are, to all seeming, passive and immobile. Yet may not this absence of motile indications be due not to want of sensibility, but to some opposing force which prevents the movement from being realised?

Imagine two persons of equal size and strength tied back to back and both struck at the same time. Each would try to bend forward and thus balance and counteract the other's movement. Yet should we be justified in regarding the two persons as insensitive to blows?

As regards the movement of Mimosa, let us for a moment think what would have happened if the upper half of the pulvinus were as sensitive as the lower. The two halves would thus have acted like opposing muscles, preventing all movement. Paradoxical as it may appear, this 'sensitiveness' of Mimosa depends on the fact that one half of its motile organ is comparatively inactive. It is as if in our human illustration we had tied a strong man to a weakling; the response of the former would then have been manifested to all eyes.

In the young stem of ordinary plants the tissue is almost equally sensitive all round, so that a contractile movement of one side is counteracted by that of the diametrically opposite side. Consequently, when symmetrically stimulated, the stem does not respond by bending to one side or the other. The whole length, however, undergoes a contraction and shortening, but to so slight an extent as to be imperceptible. This contraction of ordinary plants may, however, be recorded by means of a Mágnifying Recorder. The response of the ordinary stem is then found to be essentially similar to that of the sensitive Mimosa, though less pronounced.

The Writhing of a Tendril

The following is a striking experiment demonstrating the sensitiveness of ordinary plants. A tendril which has twined round a support is cut off and freed from the support. It appears to be altogether insensitive; but on sending a strong electric shock through it, the spiral tendril begins at once to uncurl and exhibit violent contortions, like the writhings of a worm under torture.

The explanation of this extraordinary phenomenon is as follows: the tendril, when it touches a twig, coils, as every one knows, around the support. The inner surface of the spiral becomes toughened by constant friction, while the outer side of the spiral remains relatively more sensitive. The outer or convex side of the tendril is, in fact, like the lower half of the pulvinus, and exerts a predominant contraction under excitation. The greater contraction of the convex side on electrical stimulation causes the uncurling of the tendril. Experiments such as these prove that the distinction hitherto made between 'sensitive' and 'ordinary' plants is quite unjustified; the difference between them is only one of degree.

We shall see in the next chapter that plants are profoundly affected by changes in external conditions, and that it is possible to make the plants themselves reveal those subtle internal changes, by means of responses to questioning shocks applied to them.

CHAPTER III

THE BEHAVIOUR OF PLANTS

THE diverse movements of animals are carried out by a muscular mechanism which, in all of them, serves a definite purpose. When rapid movement is needed, the reaction is quick, while in sluggish animals the response is slow. This fact is corroborated by the study of the reactions of the muscles of different animals. Thus, when we pass an electric shock through the leg-muscle of the frog, it takes a certain time to start the muscular machinery into action, there being an appreciable loss of time between the application of shock and the resulting movement. The lost time is known as the ***latent period***; in the frog, it is about $\frac{1}{100}$ of a second. The latent period is very much longer in the sluggish tortoise. On what does this difference in the rapidity of the motor mechanism depend? An attempt will be made in a later chapter to answer this question.

In the human subject also, the reaction to a shock is not instantaneous, and a short time elapses between the stimulus and the response. The duration of this latent period, moreover, is not constant at all times of the day, even with the same individual. We are more or less sluggish when getting up in the morning, become fully alert during the middle of the day, and grow fatigued and slow to respond towards the end of the working day.

Does the motile machinery of the plant exhibit the same peculiarities as that of the animal? Does the plant instantaneously respond to the shock, or is there a definite latent period? For making these very accurate measurements, our sense is not keen enough; it is, therefore, necessary to make the plant itself record these minute fractions of a second. It has been explained how the Resonant Recorder traces the response on the smoked glass plate as it slides down under the action of gravity. At a certain definite point during its descent, the moving plate establishes an electrical contact by which a shock is given to the plant. It must be remembered that, in the meanwhile, the vibrating writer is continuously tapping the time-marks. The succeeding taps in the present experiment were at intervals of $\frac{1}{200}$ part of a second. Though the plant had received the shock at the vertical mark (fig. 8),

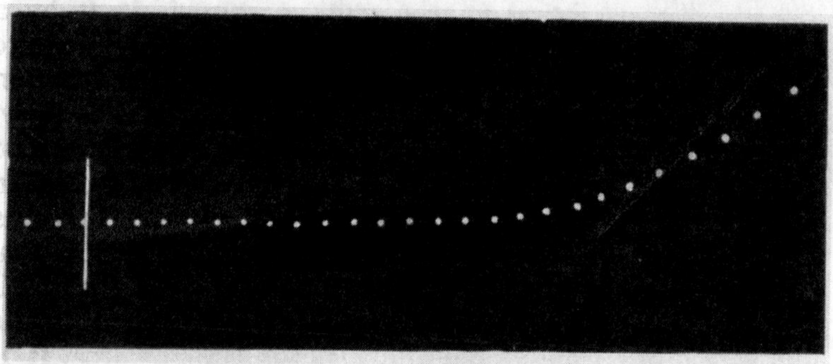

FIG. 8. Record of Latent Period of Mimosa, with 200 Vibration Recorder.

there was no immediate answer. Ten dots had already been marked, still no response. Only after the fifteenth dot did movement of the plant begin to take place. The latent period of the specimen was, therefore, 0·075 of a second.

FATIGUE

We shall next find out the effect of fatigue on the latent period. The latent period of a particular specimen, when fresh, was found to be 0·1 second. The second record was taken before the plant had recovered from its previous shock and was, therefore, in a state of fatigue. The latent period was now prolonged from 0·10 to 0·14 second. In the fresh condition, the answer is vigorous and the curve more erect. Under fatigue, the reaction is sluggish and the curve more slanting (fig. 9).

The plant thus becomes sluggish and inactive after being fatigued; when excessively fatigued, it altogether refuses to give any answer. In this dazed condition, it requires at least half an hour of complete rest to regain its equanimity.

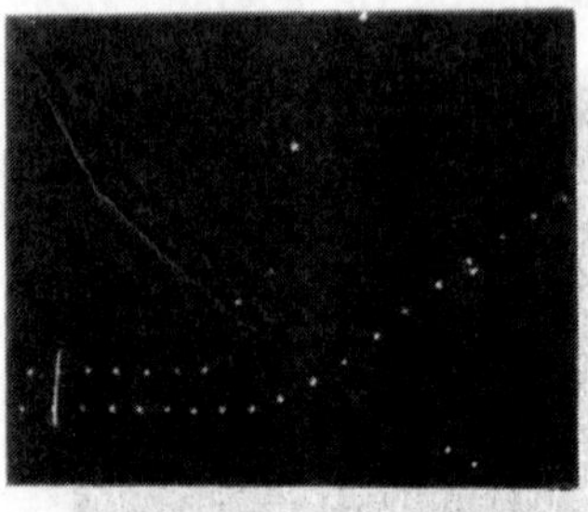

FIG. 9. Effect of Fatigue in prolonging the Latent Period. Upper record, normal; lower record, fatigue; frequency of vibration 50 per second.

There is another interesting way in which fatigue is exhibited by the plant, namely, in the reduced amplitude of its response. In this experiment, three normal responses are first obtained, allowing an interval of fifteen minutes for complete rest. The resting period is then reduced to ten minutes for the next three responses, and a depression is noticed due to fatigue. We once more allow the plant a full period of recovery, and find that its normal vigour has been regained (fig. 10).

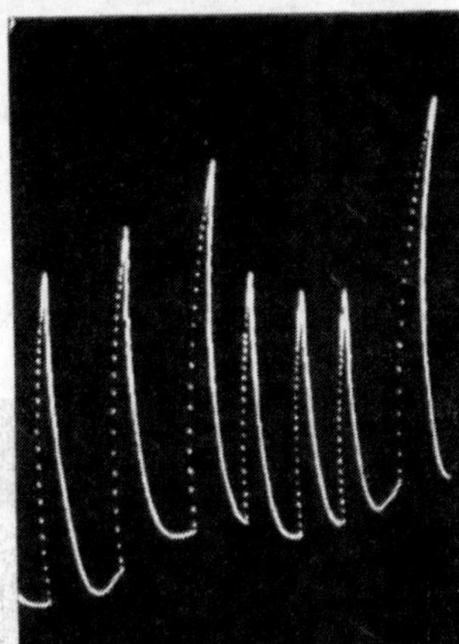

FIG. 10. Fatigue under shortened period of rest. First three uniform responses obtained at intervals of 15 minutes. The second three, with the interval of rest shortened to 10 minutes, exhibit fatigue. On returning to an interval of 15 minutes, the last record shows restoration of normal response.

In all this, we see how human-like is plant behaviour. In certain respects, the plant may even be found to be superhuman! I was awakened to this one day when taking a record of Mimosa in my laboratory. The response I was getting was uniform; but all of a sudden, there was a depression, for which I could at first discover no cause, since all the surrounding conditions appeared to be unchanged. On looking out of the window, however, I noticed a wisp of cloud passing across the sun. The plant had perceived the slight darkening which I had not noticed. As the cloud passed by, the plant recovered its normal exuberance, as the record shows (fig. 11).

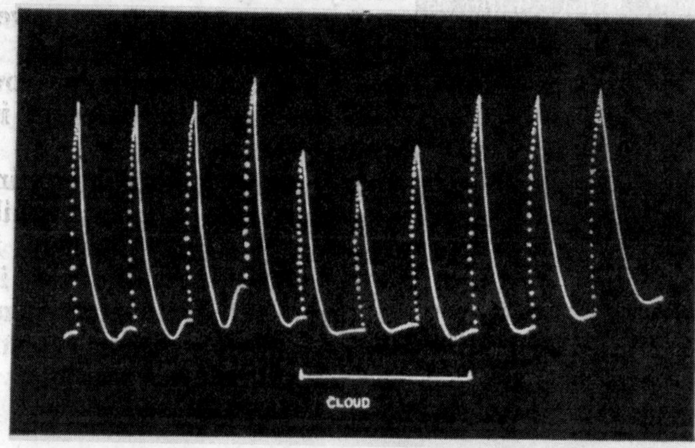

FIG. 11. Effect of passing cloud.

EFFECT OF TEMPERATURE

Not only is the plant affected by changes in light, but also by changes in temperature. There is a more or less

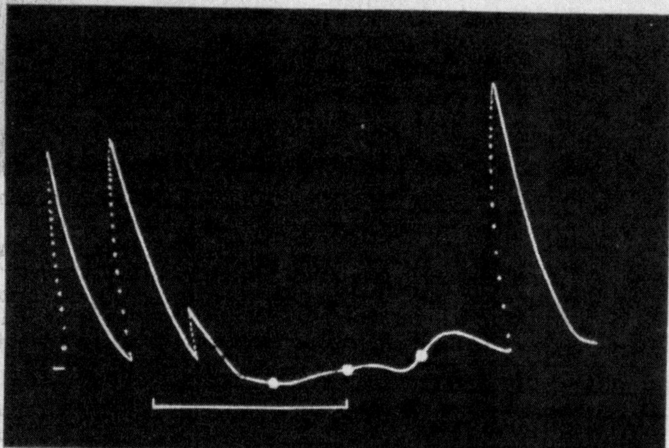

FIG. 12. Effect of exposure to intense cold.
Note sudden depression followed by abolition of excitability, also persistent after-effect.

definite temperature at which Mimosa is most active. This optimum temperature is found to be 33° C. or 91° F. When the temperature is lowered to 11° C. the plant gets numbed and more or less paralysed by cold; a few degrees of further lowering of temperature renders it quite insensitive. This is seen in fig. 12, which shows that the plant, when subjected to excessive cold, had completely lost its sensitiveness. The line below indicates the duration of the exposure to cold, after which the plant was allowed to regain the normal and favourable temperature. But the paralysing effect persisted for a considerable length of time, as seen by the lack of response to shock given at the points marked with thick dots. When the temperature is raised too high, the plant becomes oppressed with heat and its power of response undergoes a decline. At too high a temperature it gets something like a heat-stroke, and dies.

Range of Perception

Man's pretensions as a highly sensitive being receive a rude shock when certain plants are found to be a great deal more sensitive than the lord of creation. It must, however, be admitted that the plant is practically deaf, for it is little affected by sound. The case is quite different in regard to its perception of different octaves of visible and invisible light-rays. Of the multitudinous ether waves, the human retina perceives only a single octave, lying between the red and the violet. The plant not only perceives visible light, but also invisible ultra-violet and wireless infra-red waves at the two opposite ends of the spectrum.

Acuity of Perception

In human beings the most sensitive organ is the unruly tongue! Before the invention of delicate electric instruments very feeble electric currents were detected by the tongue. This can easily be verified by placing two coins, one copper and one silver, above and underneath the tip of the tongue. As soon as the two coins touch, a feeble electric current is generated by the contact of the two different metals. The irritation caused by the passage of the current gives rise to a peculiar taste. An average European can perceive by his tongue a current as feeble as six micro-amperes—a micro-ampere being one millionth part of the unit of electric current. This value appears to be subject to certain variations, depending on racial characteristics. One might expect that the tongue of the Celt would be more excitable than that of the stolid Anglo-Saxon. In any case, my Hindu students exhibit a sensitiveness twice as great as that of an average European. But the plant Biophytum is found to be eight times more sensitive than a European and four times more so than a Hindu. These revelations are as unexpected as they are startling. They show that the pretensions of man and animal to greater sensitiveness than their despised 'vegetable brethren' does not bear the test of close scrutiny.

In this general description, I have been speaking of the perception of shock by the plant, the effect of fatigue in dulling the alertness with which it gives an answering twitch, and its awareness of the slightest fluctuation of light—an awareness surpassing that of human beings.

The question naturally arises here whether consciousness should be imputed to plants. The difficulty lies in defining consciousness and in drawing a line below which consciousness does not exist, and above which it enters the domain of life.

There are two ways in which the boundary of knowledge in regard to life may be extended. We may start with the theory that the different types of living organisms have been specially created, with specific characteristics, physiological and psychological. Or, finding similarities between related forms, we may believe in gradual evolution; and thanks to Darwin's life-work, this theory has been accepted as regards the origin of different forms of life. The evolutionary process has been active not only in the development of new forms, but also in the development of special mechanisms for the performance of the various vital functions. There still exists a long prevalent idea that the physiological mechanisms of animals and plants, because they have developed along divergent lines, are fundamentally different; but the evidence adduced in this work will suffice to show that this idea is totally unfounded.

Returning to the question of consciousness, we commonly

say that our favourite animals, dogs for instance, possess it. This we realise through our sympathy; but is a fish conscious? Some say yes, and others no. In the animal kingdom, where does consciousness enter into life? Bergson discusses this problem in his 'Mind Energy':

'It by no means follows that a brain is indispensable to consciousness. The lower we go in the animal series, the more the nervous centres are simplified and separate from one another, and at last they disappear altogether, merged in the general mass of an organism with hardly any differentiation. If, then, at the top of the scale of living beings, consciousness is attached to very complicated nervous centres, must we not suppose that it accompanies the nervous system down its whole descent, and that when the nerve stuff is merged in the yet undifferentiated living matter, consciousness is still there, diffused, but not reduced to nothing? Theoretically, then, everything living might be conscious. *In principle* consciousness is coextensive with life. The amœba, when in presence of a substance which can be made food, pushes out towards it filaments able to seize and enfold foreign bodies. These pseudopodia are real organs and therefore mechanism; but they are only temporary organs created for the particular purpose. . . . It appears extremely likely that consciousness, originally immanent in all that lives, is dormant where there is no longer spontaneous movement, and awakens when life tends to free activity.'

Consciousness in animal life is thus regarded as associated with spontaneous or voluntary movement due to internal hidden causes. It is also associated with nervous reactions. These two important characteristics have been regarded as totally absent in the plant.

Leaving aside metaphysical questions on which authorities are so greatly divided, we shall deal with experimental facts and their implications. It will be shown that there is no characteristic of the higher animal which has not been foreshadowed in the plant. We shall find that all plants, even rigid trees, perceive and visibly respond to stimuli from without; that even spontaneous movement, which is so characteristic of animal life, is not absent, nor even dormant, but actively present in plant-life. I shall describe experiments in a later chapter proving that many plants possess a nervous system which shows a high degree of differentiation.

The plants, moreover, are possessed of remarkable sense-organs, by the guidance of which different parts of the body of the plant place themselves in favourable attitudes towards their environment, such adjustments reminding us of animal behaviour.

In man himself, from the moment the ovum is fertilised until he becomes fully grown, when does consciousness come into existence? Do we not see that it must be there all the time, at first latent, yet waxing imperceptibly from this latency to full blossoming? In other words, there is then a continuous evolution from rudimentary beginnings to higher perfection. Hence there is no humiliation in our kinship with the lowest of the low; rather is it a matter of pride for man to have arisen through struggle in the scale of being from a mass of protoplasmic jelly to his present state.

CHAPTER IV

DRUGGED PLANTS

THE contraction of the sensitive pulvinus of Mimosa enables us to obtain a record of mechanical response of the plant. The functional similarity between the two contractile organs, pulvinus and muscle, is not confined to the manifestation of outward movement, but can be traced to the ultimate protoplasmic mechanism.

THE TORTOISE AND THE HARE

In regard to rapidity of movement in animal muscle, the tortoise and the hare present evident differences; in the swift animal the reaction must be rapid, while in the sluggish, the response is slow. The wing-muscle of a bird of prey like the falcon is very active; that of the goose is less active, while the muscle of the domestic fowl is almost inactive, its power of flight being practically lost. What is it that confers extreme rapidity of action? Curiously enough, in the leaves of plants also there are three types of motor organ—active, semi-active, and inactive—corresponding to the three types in the animal.

MOTOR REACTION OF PULVINUS

In *Mimosa pudica* the rate of movement in response to stimulation is very rapid, the contractile fall being completed in a time as short as a second. In other sensitive plants the rate of movement is comparatively slow, as is the case with *Neptunia oleracea*, not very unlike *Mimosa pudica* in appearance. Neptunia flourishes in lakes, and in order to float in water it grows a cork belt round its stems. The movement of the leaf of Neptunia is so slow that the time taken for the leaf to complete its fall is more than a minute. Finally, the pulvinus of the Runner Bean-plant (*Phaseolus*) shows a movement which is very feeble and extremely sluggish.

Turning our attention to the motile organ, it is the contraction of the cortical cells of the pulvinus of Mimosa that produces the rapid fall of the leaf. The pulvinus of the Bean-plant is, as already stated, inactive, though its cortical cells are anatomically similar to those of Mimosa. Anatomical similarities have hitherto been unduly pressed in determining the true function of an organ; the facts given above will show the misleading character of such a criterion. It is not anatomical structure, but the protoplasmic content of its tissue that confers on an organ its physiological efficiency for discharge of a specific function. The motility of an organ is, as we shall presently find, dependent on the presence of a certain active substance in the protoplasm.

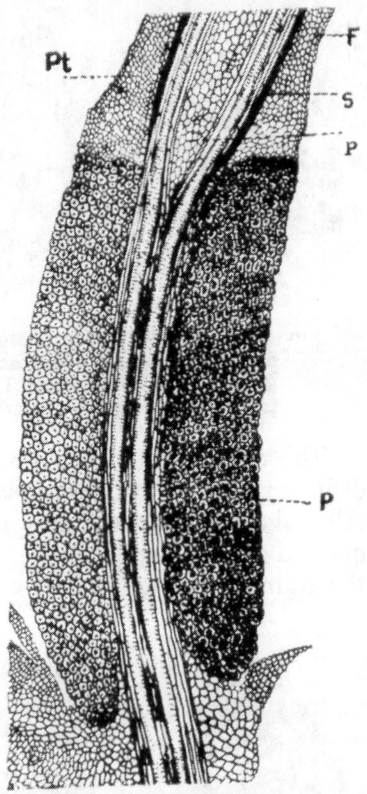

FIG. 13. Longitudinal section of petiole and pulvinus of Mimosa passing through the upper and lower vascular bundles.

Pt, the cortical cells of the petiole which remain unstained. Lower P to the right, contractile cells of pulvinus deeply stained.

DEMARCATION OF MUSCULAR TISSUE

How is the presence of an active substance to be ascertained? If we examine a longitudinal section of the pulvinus of Mimosa under the microscope, it is impossible to observe where the contractile cells begin, where they end, and how they are distributed. I succeeded, however, in distinguishing them by selective staining with safranin.

This stain produced the most remarkable results; it appeared as if a hand had, with utmost care, picked out every actively contractile cell and painted its protoplasmic contents a deep crimson (fig. 13).

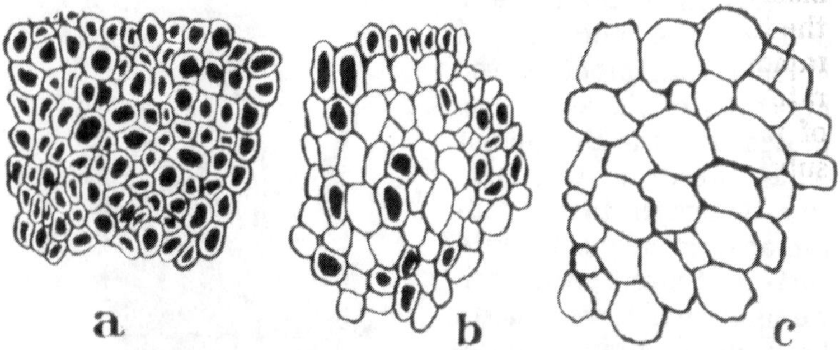

FIG. 14. Reproduction of photomicrographs of portions of the pulvini of (*a*) active Mimosa, (*b*) semi-active Neptunia, and (*c*) inactive Bean-plant. The stained content of the contractile cells represented dark.

The outline of the contractile tissue was now sharply defined; the stained protoplasmic contents of the active cells showed under high microscopic magnification a marked granular appearance. In the semi-active Neptunia, the stained contractile cells were found not to be compact

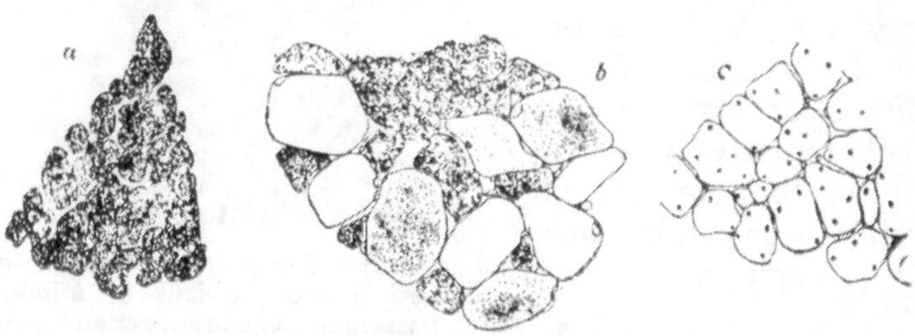

FIG. 15. Transverse sections of pectoral muscles of (*a*) the falcon, (*b*) the goose, and (*c*) the domestic fowl. The relative amount of granular substance (shown as shaded) varies directly with the bird's power of sustained flight (after Knoll).

as in Mimosa, but scattered in their distribution. No staining occurred in the pulvinar cells of the inactive Bean-plant, the active substance being here altogether absent (fig. 14). It is very remarkable that the activity of animal muscle is similarly dependent on the presence and relative distribution of an active substance (fig. 15). The pulvinus of Mimosa may thus be regarded as functionally equivalent to an active animal muscle.

All movements are ultimately dependent on oxidation or combustion. In an engine, the speed of the moving machinery is proportional to the rate of consumption of the fuel, the higher the rate of combustion, the greater the rapidity of movement. In the living machine also, the rate of oxidation plays an important part in the production of rapid movement, and in the present case the active substance is found to be highly oxidisable.

Variation of Response

We shall next make a further study of the variation of response in plants under changed external conditions. It was shown in the last chapter how animal-like, even human-like, are the general reactions of plants, and how, like us, they are affected by light and darkness, warmth and cold, rest and fatigue. There are also other factors which profoundly affect us, for example, the purity or impurity of the air we breathe. Again, there are various drugs which exert characteristic effects upon us, now beneficial, now detrimental.

Let us take, for instance, the air we breathe. The vitiated air of the town causes in us depression and lowers the general tone, whereas a trip to the country, with its ozone-laden air under the pine-trees, restores health and vigour. Think of the harmfulness of a closed stuffy room and the necessity for good ventilation. Think also of the evil habits of man, as he takes to alcohol and alcohol takes him.

Then there are the various narcotics which we take to induce sleep or to make us unconscious of the surgeon's knife. Some of these drugs, ether, for example, are practically safe; for a man may be rendered unconscious and be afterwards revived by blowing off the vapour. Chloroform is a more powerful narcotic, and though it is highly efficient, the dose may easily be exceeded with even fatal results.

Now let us enquire whether all or any of these reactions, so specific in human and animal life, have their correspondence in plant-life. We have first to find a means by which plants can be subjected to the action of different gases and narcotic vapours. For this, we enclose the plant in a glass chamber with an inlet and an outlet pipe. Gases and vapours can then be pumped into the chamber, air being expelled through the exit pipe. In order to observe the after-effect, the chamber can then be filled with fresh air. If the gases have produced only temporary narcotisation of the plant, then the introduction of fresh air revives it. But if the dose has exceeded the safety limit, or if the gas has been too poisonous, death supervenes without possibility of revival.

In the case of liquid drugs, we apply them at the roots, or at the cut end of the stem, and the suction exerted by the plant causes them to spread throughout the tissues.

Carbonic Acid Gas

I shall first describe the effect of carbonic acid gas. According to popular belief, what is death to the animal is supposed to be life for the plant. The plant is assumed to

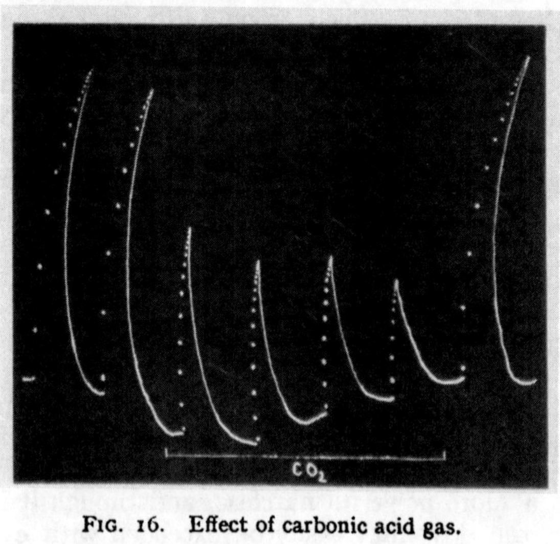

FIG. 16. Effect of carbonic acid gas.

flourish in the deadly atmosphere of carbonic acid gas. The record, however, shows that, instead of flourishing, the plant is suffocated just like a human being. Note the gasp of relief when fresh air is introduced! (fig. 16). We must remember, in this connection, that most living things, including plants, require oxygen for respiration. Unlike animals, however, green plants, in the presence of light, are able to decompose the carbonic acid gas which they take in from the air, giving off its oxygen and fixing its carbon for their nutrition. This power of carbon-assimilation is in no wise to be confused with the taking in of oxygen for respiration. The latter is an indispensable condition for the maintenance of both plant and animal life.

In contrast to the effect of carbonic acid gas is the invigorating effect of ozone, which causes an enhancement of the response.

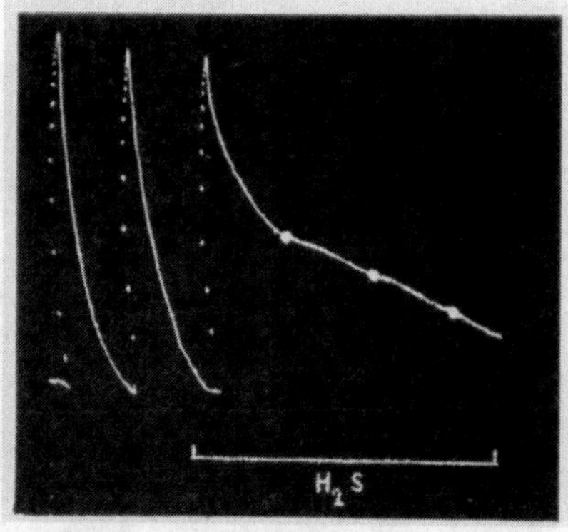

FIG. 17. Effect of sulphuretted hydrogen.

SULPHURETTED HYDROGEN

It is well known that many plants do not flourish in the atmosphere of a town. For example, the sensitive plant *Biophytum sensitivum*, in which the minute leaflets are thrown into a state of flutter by the least excitation, could not be kept in a lively condition in my laboratory in Calcutta. I found it to be fully alive and in a highly sensitive condition in the suburbs at a distance of seven miles from the city. The city air contains traces of various gases, such as sulphur dioxide and sulphuretted hydrogen, both of which are highly injurious to the plant. This is seen in fig. 17. Here the introduction of sulphuretted hydrogen abolished all sensibility; the three thick dots represent the application of strong shocks, to which there was no response. The plant was afterwards found to be dead.

Ether

From such deadly poisons we turn to narcotics. First we try ether. The record (fig. 18) shows that after an application the plant begins to lose its excitability, just as a person loses consciousness after etherisation. On blowing off the narcotic vapour, the plant is seen gradually to regain its normal sensibility.

The results of other experiments show that a very minute dose of ether causes an increase of excitability. Here we come across the remarkable fact that the effect of a drug is changed by the amount of the dose applied ; a minute

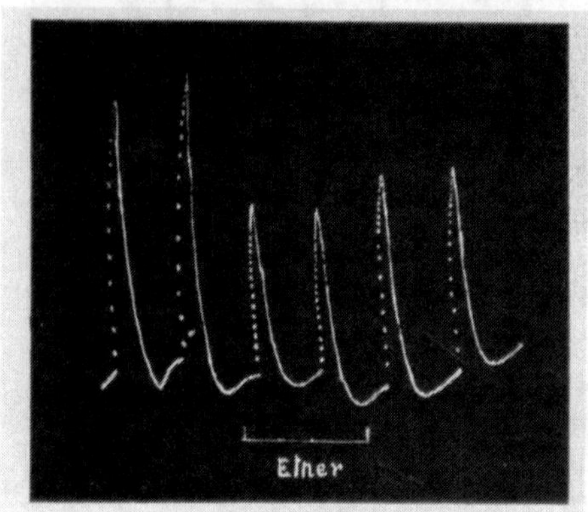

FIG. 18. Effect of ether.

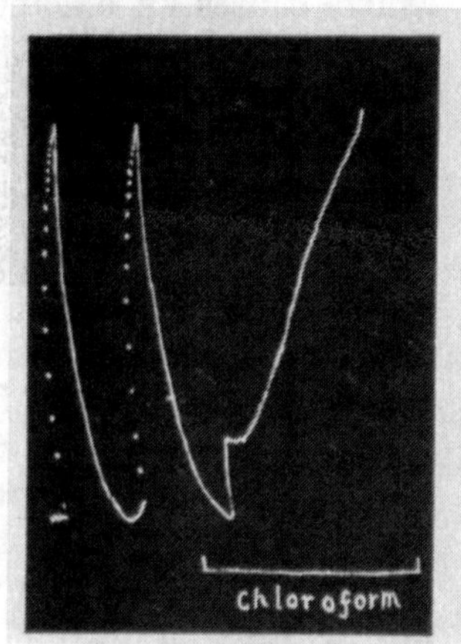

FIG. 19. Effect of chloroform.

dose, generally speaking, produces a result diametrically opposite to that of a larger dose.

Chloroform

Ether is found in medical practice to be a safer narcotic than chloroform, any excess of which is attended by fatal results. In fig. 19 is shown the effect of a large dose of chloroform on the plant. This not only produced a total abolition of excitability, but brought about a sudden spasm which is seen as a line shooting upwards. After this, the blowing off of the vapour failed to revive the plant, which showed the characteristic discoloration of death.

ALCOHOL

The immediate effect of dilute vapour of alcohol is often to produce a transient enhancement of excitability. But a depression results from its continued action.

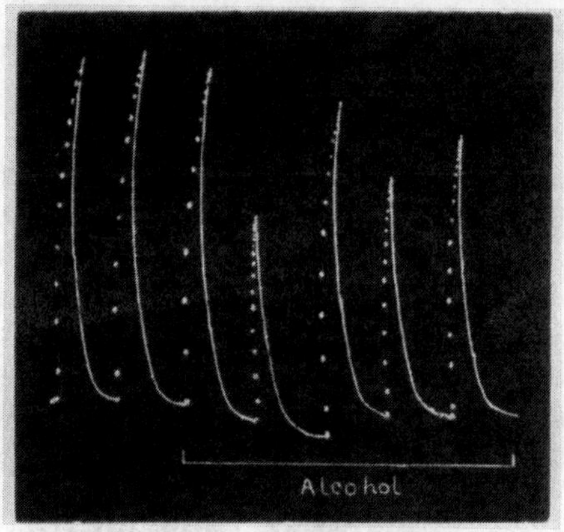

FIG. 20. Effect of alcohol.

The ludicrously unsteady gait of the plant under intoxication (fig. 20) could, no doubt, be effectively exploited in a temperance lecture!

CHAPTER V

THE ELECTRICAL RESPONSE

A LIVING tissue is said to be sensitive because it responds to stimulation from outside. As long as it is alive, so long will it respond and then recover, making itself once more ready for a new response. The brief disturbance of the living poise, to be restored to equilibrium of itself, is quite unlike the rolling of a stone downhill after a push. For the stone cannot of itself regain its original position ; but the living tissue quickly reasserts its stable poise on the cessation of the stimulation.

The response of the plant to stimulation, as recorded in the foregoing chapters, has been that of mechanical movement effected by the contraction of a motile organ. The intensity of the response was found to depend on the presence of a certain active substance in the cells of the organ. Vegetable tissues in general are, however, physically restrained from exhibiting that responsive movement which we have hitherto accepted as a test of the sensibility of a tissue. The very important question arises as to whether a tissue which cannot show any movement is nevertheless sensitive to stimulation ; whether all plants and their different organs, leaf stem, root, flower and fruit, are irritable. In order to answer this question, it is necessary to devise a method by which excitation in living organs may be detected even in the absence of movement.

Movement is not the only manifestation of excitatory reaction. Stimulation induces an electrical change in an irritable tissue. If we take a piece of living stem, any two points on its surface, A and B, will be found to be in the same electrical condition, provided both are in a state of rest. The similarity of electrical condition at the two points A and B may be demonstrated by suitably connecting them with a galvanometer, which is a sensitive detector of electrical current. When the two points are at rest, they are in the same electrical condition, and therefore no current flows through the galvanometer. If next we excite the point B, say by pinching or striking it, then the electrical level of B in relation to A will be disturbed, and an electrical current will be found to flow from A to B through the galvanometer, which will show a deflection. The excited point B will be galvanometrically negative with respect to A. Stimulation of A gives rise to a current in the opposite direction (fig. 21). For the sake of simplicity, we shall henceforth describe the electrical change of the excited point as negative. After a time, as the tissue recovers, the electric change, due to excitation, will also disappear.

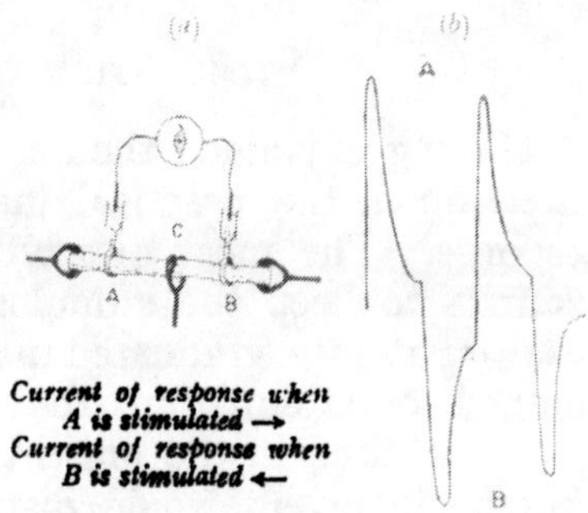

FIG. 21. (*a*) Method of obtaining the electric response.
(*b*) Stimulation at A produces a current of response from B to A across the galvanometer (up-curve); stimulation at B gives rise to a current in reverse direction (down-curve).

The following method was adopted to record the induced electrical changes. The galvanometer which is to record the responsive electric current consists of a coil of wire in the centre of which is suspended a small magnet with its length parallel to the plane of the coil. When a current flows round the coil, the magnet rotates in one direction or another, depending on the direction of the current. A small mirror is attached to the suspended magnet, and a spot of light reflected from the mirror magnifies the rotation of the magnet. The reflected ray of light acts as a long magnifying index without weight. For inscribing the record we use a sensitive plate, and the moving ray of light photographically records the response and recovery. We shall presently find that the electrical pulse of response is a faithful indicator of the vitality of a tissue, the electrical response disappearing at the death of the tissue.

Before proceeding further, it is desirable to get a clear conception of the various manifestations of excitation or its opposite. Under normal conditions, the tissue of the plant is tense or turgid with sap, it is in a state of turgor. After stimulation various symptoms appear as so many signs of excitation. These are (1) contraction, (2) diminution of turgor, (3) movement such as the fall of the leaf in Mimosa, and (4) negative electric change. The opposite process, which brings about recovery, and may even carry it to excess of recovery, is characterised by (1) expansion, (2) increase of turgor, (3) movement of erection as in leaf of Mimosa, and (4) positive electrical change.

Stimulation of the Tissue

Having explained the general principles, we shall next enter upon the practical method of obtaining electrical response. The most difficult problem in stimulating the plant is to keep the stimulus constant, or to increase or decrease it in a graduated manner. The most satisfactory method of stimulation, that of an electric shock, could not be utilised in the present case, because the leakage of the shock-current would disturb the current of response. Hence a non-electrical method of stimulation by torsional vibration had to be devised.

If we hold our finger and cause a slow twisting movement, the stimulation produced will be slight; but if the torsion be made very suddenly, the stimulus will be intense, and cause a painful sensation. The efficiency of a stimulus is, in general, found to depend on the abruptness of its impact.

In experimenting with plants, we take a piece of stem, A B, and hold it in the middle by a vice, C, which acts as a block, preventing excitation of one half from passing over into the other. One of the ends, say B, is held in three clamping jaws H. A torsional vibration is now imparted to this end by the handle K. The angle of torsional vibration, which determines the intensity of stimulus, can be accurately measured by the graduated circle and may be predetermined by a sliding stop. Electrical connections are made with A and B through E, E′, which lead to the galvanometer. The plant is placed in a glass chamber in which it can be subjected to the action of high temperature by means of heating the electric coil R. For experiments on the action of anæsthetics, the vapour can be blown through the side tube (fig. 22).

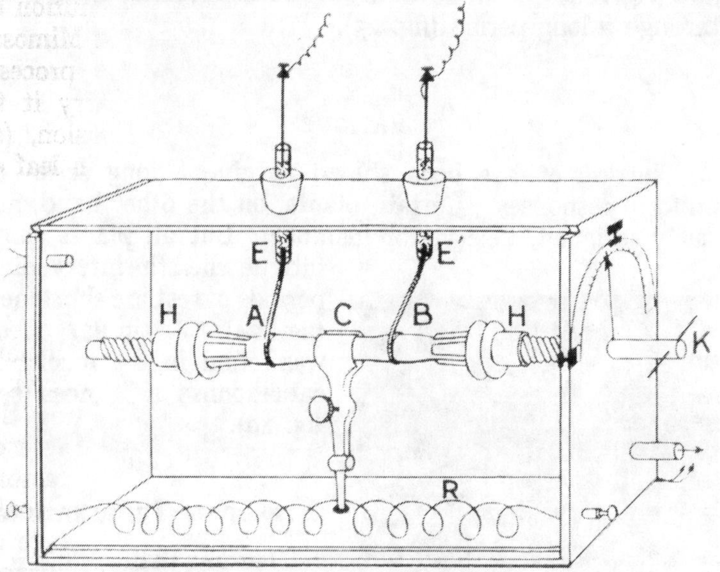

Fig. 22. Apparatus for observing modification of electric response under external variation.

Uniform Response

I shall now give records of the electric response of the plant and its modification under varied external conditions. As regards sensitiveness in ordinary plants, we cannot

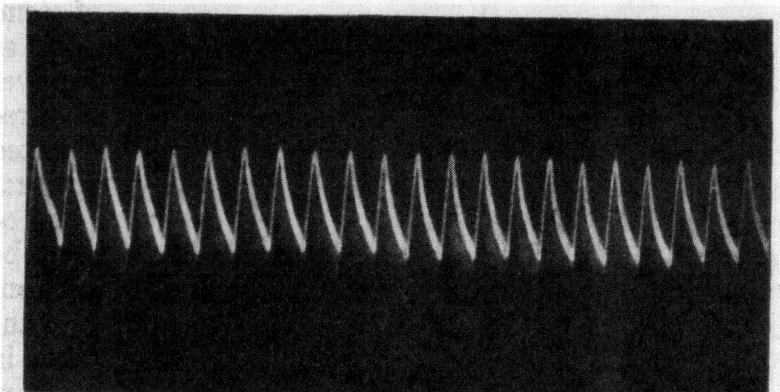

Fig. 23. Uniform electric responses under torsional stimulus (Carrot).

imagine anything more stolid and undemonstrative than a carrot. It is a revelation to find how excitable it is, and how vigorous and uniform are its successive responses even through a long period (fig. 23).

Fatigue

The Carrot has been shown to give a long series of uniform responses. Certain plants, on the other hand, are easily fatigued, Celery, for instance. But all plants show fatigue when the intervening periods of rest are shortened (fig. 24). A similar effect was found in the mechanical response of Mimosa (see fig. 10).

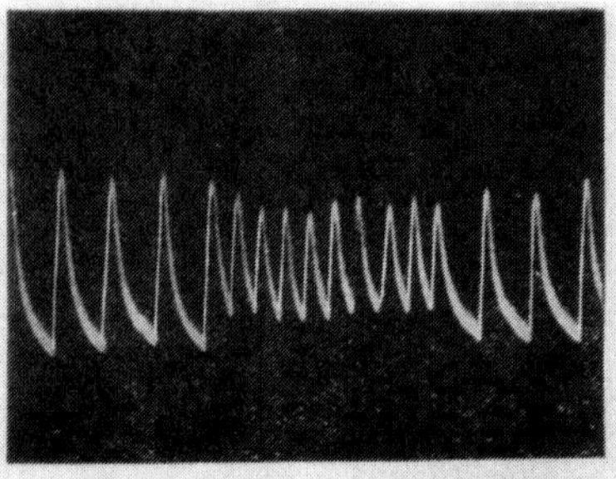

Fig. 24. Fatigue of electric response under shortened period of rest.

Effect of Temperature

Turning to the minimum temperature which is fatal, tropical plants succumb to it earlier. Thus the tropical Eucharis lily, when subjected to a freezing temperature for fifteen minutes, had its electrical response completely abolished, whereas northern plants like Holly and Ivy, when subjected to the same temperature, continued to give their electric response.

Effect of Poison

I took twenty leaf-stalks of Horse-chestnut and divided them into two batches of ten each. One batch was kept in water to serve as control, and the second batch had the cut ends placed in solution of mercuric chloride, a well-known poison. The plants had numerous plant-lice living on them. It was observed after a period of twenty-four hours that the average electric response of the leaf-stalks standing in water was twenty-three divisions of the galvanometric scale, and that the lice on them were still living.

On the other hand the leaf-stalks that had been standing in solution of mercuric chloride gave no electric response on stimulation. Evidently they had been killed by the poison, a conclusion strengthened by the fact that the lice upon them were dead.

Death under Chloroform

In the cases given above, it was only the last fatal phase that was observed. It appeared important, how-

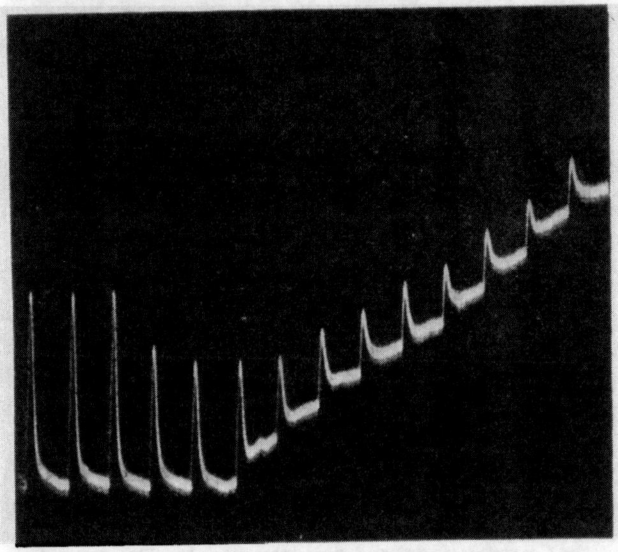

Fig. 25. Gradual abolition of response under chloroform.

ever, to trace the gradual effect of the administration of a poison culminating in death. The method adopted was to obtain first a series of normal responses to uniform stimulation; subsequently, without interrupting the procedure, the poisonous agent, chloroform vapour, was blown into the plant-chamber.

It is seen in fig. 25 how the anæsthetic gradually abolished excitability, as indicated by the increasing depression of the response. Long-continued action of the narcotic ultimately proved fatal and produced a permanent abolition of electric response.

Plant Scalded to Death

These facts suffice to show that the modification of vital activity can be accurately gauged by means of the electric response, which gives the clearest indication of the depression or abolition of vitality. This is strikingly shown in the following record of a case in which the plant was

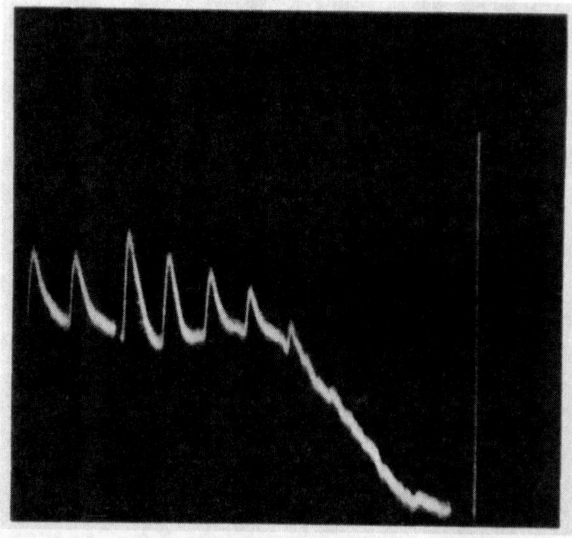

FIG. 26. Abolition of response by steam.

The first two responses (left) took place under a temperature of 17° C. Steam was then admitted. At the end of 5 minutes the Carrot was dead and response ceased. The vertical line (right) represents 0·1 volt.

gradually scalded to death by the introduction of steam into the plant-chamber. The first two records were taken at the temperature of the room, which was 20° C. Steam was gradually introduced; this, at first, caused a transient excitation, after which there was a rapid decline, and the response was completely abolished at death by scalding (fig. 26).

Electro-motility, that is to say the power of exhibiting electro-motive change under stimulation, is the test for the vital condition of a tissue. Similar reactions have long been known in animal tissues, and now we have plants reacting in the same way. By the employment of this test, we are able to prove that every plant and every organ of the plant is sensitive; they all answer to stimulation by a definite electric response. It has also been shown that in the modification of the electric response by fatigue, by high and low temperature, by the action of narcotics and poisons, the phenomena strictly correspond to similar reactions of animal tissues.

In this and in preceding chapters the responses of the plant to stimulation have been discussed. The conclusions to be drawn are that all plants give a response of negative electric variation, and that certain plants possessing a special motile mechanism also give a mechanical response of movement; the two forms of response are completely concurrent in all respects.

CHAPTER VI

THE SLEEP OF PLANTS

THE leaflets of many plants are sensitive to light; they become completely closed at the approach of darkness, and open again when light returns. Misled by the similarity to the closure of our eyes in sleep, the closing of the leaflets was taken to signify sleep in the plant. The resemblance is merely fanciful, for a similar closure is often found to take place at midday under the stimulation of strong light.

Many involuntary activities of life show no sign of cessation, but persist both day and night; in the animal, the heart continues to beat, the processes of digestion and of building up of the tissues continue without intermission. Similarly in the plant, circulation of food-material continues for the supply of the local wants of organs overworked by day and the accumulation of material for tomorrow's needs; growth also persists, increasing the bulk of the organism and making good its injuries. Life is sleepless activity.

Yet we are periodically switched off, and then become unconscious of all external happenings, and pass through a daily cycle of what we know as sleeping and waking. The question as to whether plants sleep or not can be put into the form of a definite inquiry: is the plant equally aware of external happenings throughout day and night? If not, is there any period when it loses its perceptive power? Is there, again, another period when it wakes up as it were to a condition of maximum alertness?

How are we to find whether the plant really sleeps or not? When we wish to know if our friend is asleep, we shake him and ask, 'Are you awake?' If he is, he shouts, 'Yes.' If he is half asleep, he gives some indistinct answer. But when he is fast asleep, he gives no answer. It is only when his sleep is over that he awakes and gives his normal and vigorous answer.

A somewhat similar procedure has to be adopted in the case of the plant. We take the sensitive Mimosa and question it hour after hour: 'Are you awake?' 'Are you awake?' 'Are you awake?' and the changing vigour of

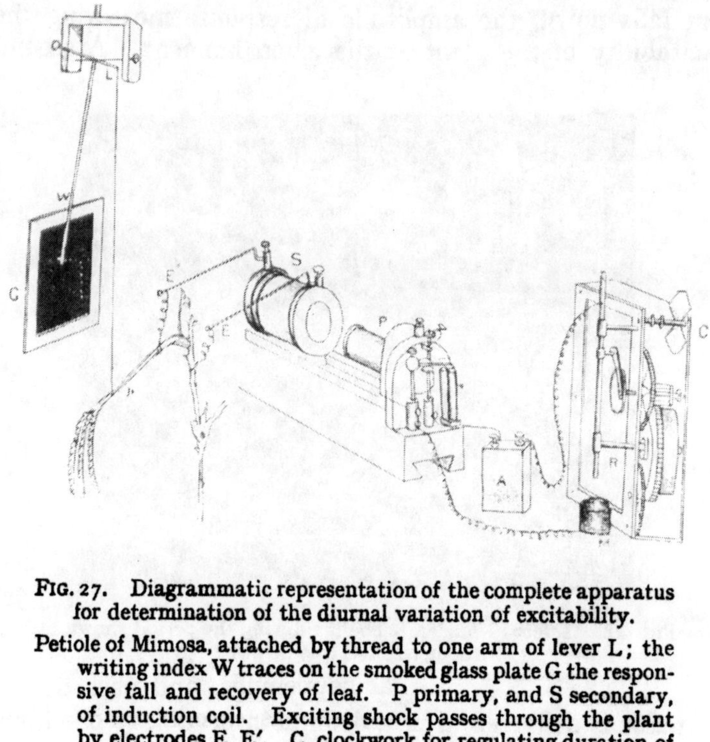

FIG. 27. Diagrammatic representation of the complete apparatus for determination of the diurnal variation of excitability.

Petiole of Mimosa, attached by thread to one arm of lever L; the writing index W traces on the smoked glass plate G the responsive fall and recovery of leaf. P primary, and S secondary, of induction coil. Exciting shock passes through the plant by electrodes E, E′. C, clockwork for regulating duration of shock. Primary circuit of coil completed by plunging rod, R, dipping into cup of mercury, M.

the plant's answer at different times indicates if and how much it is asleep or awake.

The difficulty of the problem lies in delivering the same questioning shock every hour of the day and night, and recording the plant's answer in the form of a responsive movement. It was solved by the special Sleep-Recorder devised for the purpose, represented diagrammatically in fig. 27.

The Sleep-Recorder

The leaf of Mimosa is attached to one arm of an aluminium lever L by means of a thread; at right angles to the lever is the writer W, which traces the responsive movement of the leaf on a smoked glass plate, which is allowed to fall at a definite rate. Under a definite stimulus the leaf falls down, the amplitude of response measuring the excitability of the plant at the given moment. A second

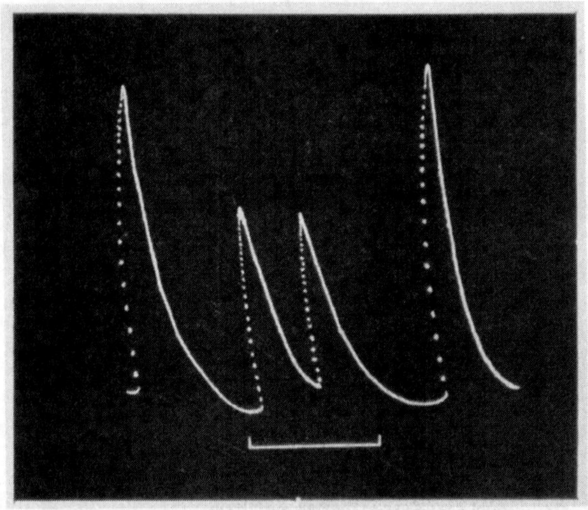

Fig. 28. Effect of moderate cooling during the period shown by the horizontal line below.

stimulus is delivered after a definite interval, say one hour, and the corresponding response shows whether the excitability or awareness of the plant has remained constant or has undergone any variation.

The questioning shock is electrical, the effective intensity of which has to be maintained absolutely uniform. The secondary coil is placed at a suitable distance from the primary, and the duration of the shock is kept constant by completing the primary electric circuit for a definite length of time. The completion of the primary circuit is made by the plunging-rod R, which dips into a cup of mercury for only 0·2 second. The record of the response is obtained on the smoked plate, which is kept oscillating to and fro by clockwork.

It will presently be shown that the plant goes through a daily cycle of sensibility and insensibility which may be aptly described as waking and sleeping. The cause of this has been fully analysed elsewhere.[1] I may briefly enumerate some of the factors which cause sleep or insensibility of the plant. The plant becomes insensitive under prolonged darkness; lowering of temperature has even a more pronounced effect in depressing the excitability (fig. 28). It should be remembered that the plant is subjected, during a complete day, to considerable variations of temperature, the minimum being early in the morning and the maximum at about noon.

Response at Middle of the Day

The 'awareness' of the plant is at its maximum throughout the middle of the day. This will be seen in

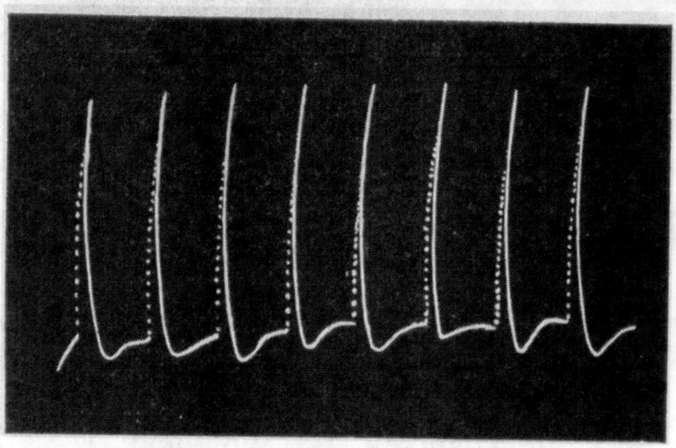

FIG. 29. Midday record from noon to 3 P.M., exhibiting uniform and maximum excitability.

the record reproduced (fig. 29), which show how vigorous and uniform are its responses. This period may be regarded as the 'office hour.'

[1] Bose—"The Diurnal Variation of Moto-Excitability in Mimosa"—*Annals of Botany*, Oct. 1913.

Its wakefulness shows a striking change in the morning and evening.

WAKING IN THE MORNING

Mimosa is a late riser; this will be seen more clearly in the complete record for twenty-four hours. I reproduce the record (fig. 30) taken every half an hour, illustrating

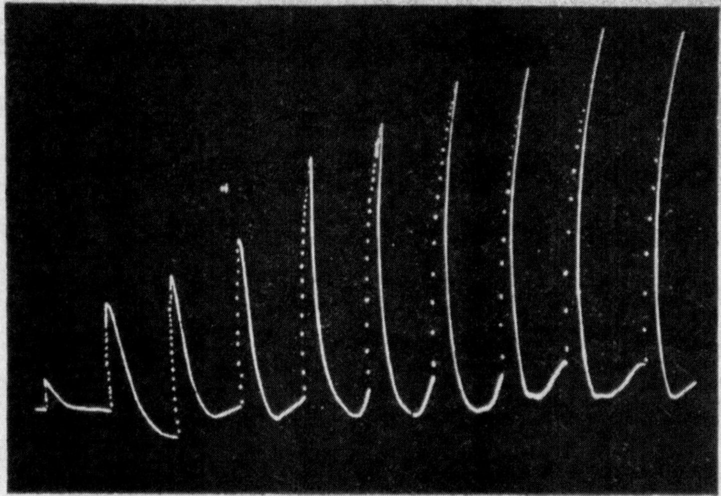

FIG. 30. The gradual waking up of the plant from 8 to 12 A.M.

the process of slowly awakening at about 9 A.M. Note the human way in which it is trying to rouse itself gradually.

RECORD FOR DAY AND NIGHT

This record was commenced at 5 P.M. and taken every hour of day and night for twenty-four hours. The plant was fully awake for the first two hours. It became somewhat sleepy between 9 P.M. and 2 A.M., but was still awake at 6 A.M., after which it fell sound asleep, and remained so until past 8 A.M. (fig. 31).

Rousseau told us that modern life is decadent and that the only hope of salvation was a return to primitive life.

Surely we could not go back to anything more primitive and unspoilt than unsophisticated plant-life. And yet

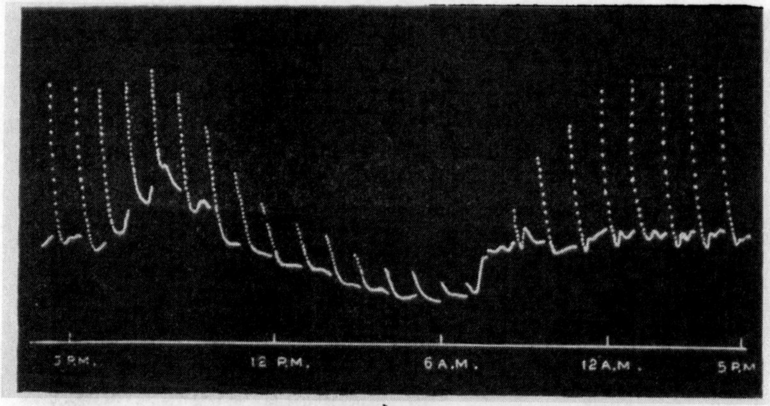

Fig. 31. Diurnal record showing variation of sensibility of Mimosa from 5 P.M. to 5 P.M. next day.

what are the facts? Apparently Mimosa has forestalled the dissipated life of the modern Babylons of London, Paris, and New York, keeping up all night and going to bed only after the rising of the sun!

CHAPTER VII

THE PRAYING PALM OF FARIDPORE

As if in protest against the imputation of leading a fast life, some plants assume an appearance of a highly sanctimonious character. At least, that was the impression made on the credulous public by the remarkable performance of the Praying Palm. While the temple bells rang, calling the people to prayer in the evening, this tree bowed down its head in adoration. By the morning its head was again erect, and this process was repeated every day of the year. This extraordinary phenomenon was regarded as miraculous, and pilgrims were attracted in large numbers. It was alleged that offerings made to the tree had been the means of effecting marvellous cures. It is not necessary to pronounce any opinion on this subject; these cures may have been effective in the same way as other faith cures now prevalent in the West.

This particular Date-Palm (*Phoenix sylvestris*) was a full-grown rigid tree, its trunk being about 10 inches in diameter. It must have been displaced by a storm, so that the trunk had an average inclination of about 60° to the vertical. In the course of its daily movement the trunk, throughout its entire length, was elevated in the morning and depressed in the afternoon: the upper end of the trunk thus moved up and down through a distance of 3 feet. The 'neck,' or upper end of the trunk, bearing the leaves, was concave to the sky in the morning; in the afternoon the curvature was reversed. The large and long leaves which pointed high up against the sky in the morning were thus swung round in the afternoon through a vertical distance of about 15 feet.

To the popular imagination the tree was a living giant, rearing its head in the morning to more than twice the height of a human being, leaning forward in the evening from this towering height, and bending its neck till the head of the leaves pressed against the ground in an attitude of devotion.

I took two photographs of the tree, one in the morning and the other late in the afternoon (fig. 32). I was

FIG. 32. The 'Praying' Palm.

The upper photograph represents the morning, and the lower photograph the afternoon position.

completely puzzled by the peculiar behaviour of the tree, the like of which had never been noticed before. Natural science does not believe in the occult, for to it nothing is extra-physical, but only mysterious owing to some hitherto unascertained cause.

What could be the cause or causes behind the mystery? The phenomenon repeated itself day after day, and the only changing factors were light and temperature. Could it be light? This was not likely, for in order that it could act, light would have to come into contact with the living tissues. But the thick bases of dead leaves completely screened the deep-lying living tissues from light. The only other factor was the changing temperature, which, by affecting growth, causes movement, as illustrated by the movement of the Water-Lily described in a later chapter. But this Palm-tree was old and not in a state of active growth.

The only means of solving the problem was the construction of a special apparatus which would record the movement of the tree continuously throughout the day and night: the up-movement was recorded as an up-curve and the down-movement as a descending one. The changing temperature was also recorded simultaneously on the same recording plate by means of a metallic thermometer, the rise of temperature being indicated by the fall of the thermal curve, and *vice versa.* The principle of the recording apparatus will be understood from the illustration given later (see fig. 34).

A difficulty arose at the beginning in obtaining the sanction of the proprietor to attach the recorder to the tree. He was apprehensive that the miraculous power of the tree might disappear under the profane contact of foreign-looking instruments. His misgivings were removed on the assurance that the instrument was made in my laboratory in India, and that it would be attached to the tree by my assistant who was the son of a priest.

The results obtained were most striking. A casual observation had led people to believe that the tree lifted itself at sunrise and prostrated itself at sunset, but the continuous record obtained showed that the tree was never at rest, but in a state of continuous movement, which underwent periodic reversal (fig. 33). The movement was

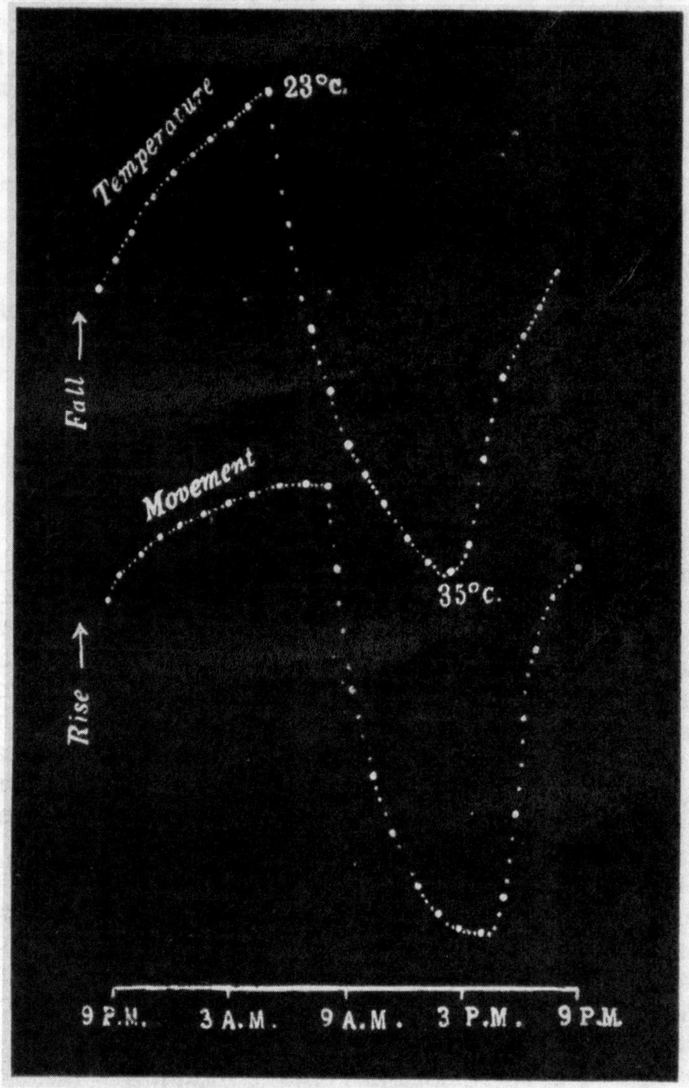

FIG. 33. Record of diurnal movement of the 'Praying' Palm (*Phoenix sylvestris*).

Thermographic curve for 24 hours, commencing at 9 in the evening, is given in the upper record, in which the fall of temperature is represented by an up-curve. The corresponding diurnal curve of movement of the tree is given in the lower. Successive dots mark intervals of 15 minutes; thick dots intervals of an hour.

not passive, but was effected with an active force sufficient to lift a man off the ground.

The tree attained its maximum erection at seven in the morning, after which there was a rapid fall. The downward movement reached its maximum at 3.15 P.M., after which it began to lift itself very slowly, then more rapidly, until, as stated before, it lifted itself to the highest position at seven next morning.

The accurate record made by the tree itself showed that the maximum fall did not take place exactly at prayer time; but people who are out to impute miracles would not be disturbed in their faith by such a slight discrepancy.

Discovery of the Cause of Movement

The possible explanation of the daily movement was some periodically changing condition, such as light or temperature. Light, as we saw, could not be the effective cause. Closer inspection of the curve of daily movement also supported this contention. For if the movement were due to light, then there would be a climax at noon and the opposite at midnight. But instead of this we found that the highest point was reached not at noon but at seven in the morning. The lowest point, on the other hand, was attained not at midnight but in the afternoon.

Turning next to the variation of temperature, we were at once struck by the fact that the curve of the movement of the tree was practically a replica of the curve of the variation of temperature. The rise of the tree was seen to follow the fall of temperature, and *vice versa*. The movement of the tree lagged behind the change of temperature, and the delay was due to two causes: it took some time for the thick trunk of the tree to attain the temperature of its surroundings; in addition to this physiological inertia delayed the reaction.

Physical or Physiological?

Having proved that the movement was caused by variation of temperature, it now remained to be seen whether this was due to mere physical expansion and contraction, or to some specific reaction of life. The conclusive test would be to observe what happened after the death of the tree. In such a case, the physical movement would persist, while the physiological action would disappear. In order to decide this question it was, of course, impossible to take liberties with the life of a tree which was so unique and held in such reverence.

But a year afterwards, while presiding at my lecture on this subject, Lord Ronaldshay, the Governor of Bengal, announced that a telegram had just reached him from his officer at Faridpore stating that 'the palm-tree is dead and its movements have ceased.' This much lamented demise of the tree offered the means of solving the problem. It was unfortunate that the tree had to die in order to prove that its movements had been due to its vital activity.

The superstitious might connect the death of the tree with the liberty I took of recording its movements. Needless to say, the two had no causal relation ; the tree was old and died a natural death more than a year after my experiments. Nevertheless I felt uncomfortable about it and wished the matter to be buried in oblivion. But this was not to be for a considerable time, since the Press took up the matter and made comments, of which the following is a specimen :

'*Tragedy in high life!* The praying palm of Faridpore, as already reported, expired suddenly on the very eve of Dr. Bose's lecture. It must have heard that the distinguished scientist was about to expose a humbug. That pious attitude, we are told, was entirely due to the effort to keep itself warm ! At the same time, we cannot help thinking that it may have been a sign of disease, for how is it that other palms never " pray " ? '

For once the omniscience of the Press was at fault. For not only the sacred tree of Faridpore but other Palms and even common trees have this habit of 'praying.' I have received information of another Palm growing by the side of a pool, the trunk of the tree being inclined towards it. The uplifted leaves of this tree are swung round in the afternoon and dipped into the water of the pool.

The phenomena described above are not marvels of the mystical East; a similar thing happened in the prosaic neighbourhood of Liverpool. An English friend sent me the following extract from the *Liverpool Mercury*, dated December 13, 1811:

'*Remarkable Phenomenon:* There is at present a willow tree of considerable height and about three yards in circumference, growing on the banks of a rivulet on the farm called Tubsill, the property of the Rev. Mr. Wasney, near Shipton, which actually appears *animated*; it will at times prostrate itself at full length on the ground, and then rise to its original perpendicular position. Incredible as this may appear, it is a fact and has been the astonishment of hundreds who have seen it!'

Another incident has recently been related to me by a coco-nut planter in South Africa. Owing to the strong prevailing winds from the sea, the rows of Palm-trees on his plantation had grown slantingly. It had been a perpetual wonder to him that while in the morning all the Palm-trees were fully erect, so that the coco-nuts were beyond reach, in the afternoon the trees were so accommodating as to bend down to make plucking easy!

Trees growing at even a slight inclination show characteristic up and down movements; these are so slow that a casual observer fails to notice them. The movements traced by the automatic recorders, however, are very marked and definite.

The movements of trees like the Faridpore Palm become conspicuous for two reasons. First, the geotropic action, on the variation of which the movements depend, is most intense when the tree is nearly horizontal, and less so when more erect. Moreover, the movement of an inclined tree is more noticeable, since the ground affords a fixed object for reference. With a more erect tree, the movement would escape notice, since it would be executed with empty space as a background.

Explanation of the Phenomenon

The movement of the floral leaves of the Water-Lily will be shown in a subsequent chapter to be brought about by variation of growth, induced by changes of temperature. In the case of fully grown rigid trees, it is necessary to seek another explanation. I cannot give here an account of a large number of experiments which led to the discovery of a class of phenomena hitherto unsuspected, namely, the influence of temperature on geotropic reaction.

When a plant-organ, whether stem, branch, or leaf, grows slantingly, the stimulus of gravity tends to raise it upwards. Lay a potted plant horizontally on the ground, and it will soon be noticed that the stem and leaves raise themselves upwards against the pull of the earth. The geotropic reaction has, therefore, to overcome the tension of the plant's tissues. Equilibrium is attained when these counteracting forces balance each other. The experiments carried out show that the geotropic reaction is modified by change of temperature; a rise of temperature diminishes, while a fall accentuates the geotropic up-curvature. Hence the 'dynamic balance' is upset in one direction or the other by the rise or fall of temperature, thus causing the daily up and down movements.

Theoretically, all inclined organs sensitive to the stimulus of gravity should exhibit the characteristic movements under variation of temperature. Now the horizontally outspread leaves of plants are also subject to the stimulus of gravity. Do they too exhibit daily movements similar to those of the Praying Palm?

The Automatic Recorder

This investigation was carried out with the help of an Automatic Recorder specially constructed for the purpose. The apparatus has four recording levers: the first three record movements of leaves (or of shoots horizontally laid) of different plants; the fourth lever records the variation of the temperature by means of a metallic thermometer (fig. 34). The smoked glass plate is made to oscillate to and fro by clockwork at intervals, say, of fifteen minutes. Four records are thus made at the same time, the uppermost being the thermograph, and the three others the phytographs of different leaves.

The Leaf-Autographs

The records show that the leaves are in a state of continuous movement and that they all exhibit a rise during

Fig. 34. The Leaf-Recorder.

a fall of temperature, and *vice versa*. There is, however, an individuality of each species of plant, exhibited in its characteristic autograph (fig. 35). Note the tremulous writing of Papaya; the signature of Croton shows more character, it evidently crosses its 't's' with greater decision!

It is thus possible to identify a plant from its peculiar handwriting.

The plant-world, as a whole, is thus not passive but fully responsive. The full-grown and apparently rigid tree

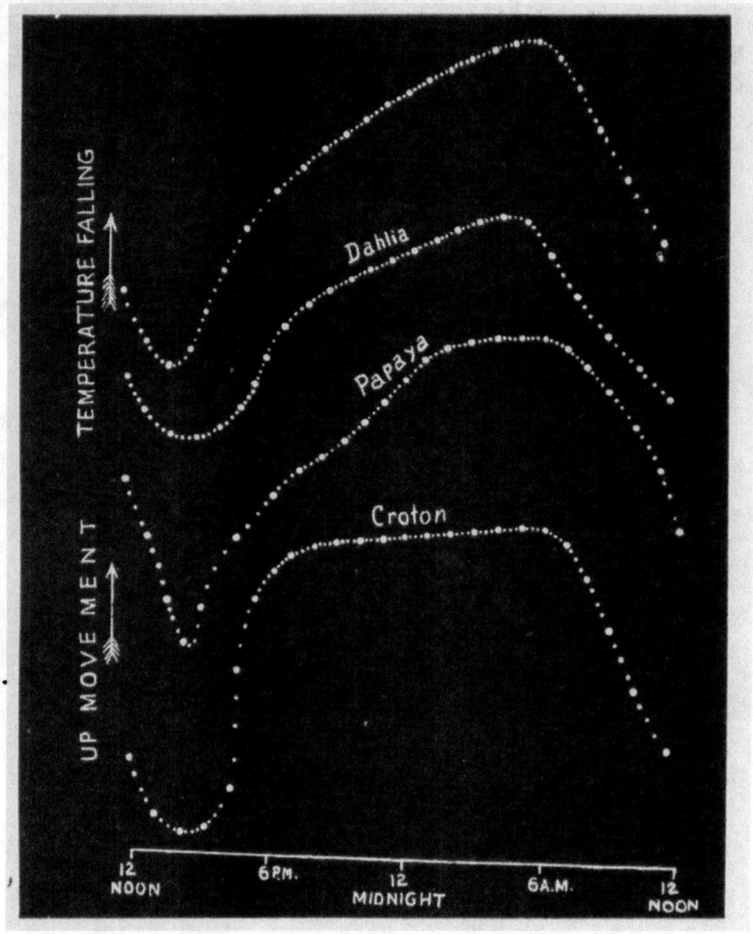

FIG. 35. The Autographs of Dahlia, Papaya and Croton.

is sensitive to the slightest external change, even to the passage of a cloud across the sky, and it signals its perception of the change by a definite movement. The arbitrary distinction between 'ordinary' and 'sensitive' plants thus wholly disappears: it is a question of degree only. The facts show that not only the 'Praying Palm' but every tree and its different organs perceive changes in the environment, and execute movements in response to them. It is not the cushioned pulvinus alone that is excitable, but the tree as a whole is instinct with sensibility. Its seemingly rigid trunk is really a gigantic pulvinoid which perceives and responds to the multitudinous stimuli of its environment.

CHAPTER VIII

RESPONSE OF INORGANIC MATTER

THE results given in the previous chapters show that a continuity exists between the reactions of animal and vegetable organisms, proving that the laws of life are one. In studying this continuity, some are apt to make the mistake of comparing a man with an amœba, and wonder that all the complicated organs in the highest are not found in the lowest. The fact of the matter is that there has been a continuous evolution from the lowest to the highest, with corresponding increase in complexity.

Simple machinery does not easily get out of order, and in the machinery of life we find the simplest mechanism not easily deranged. Hence unicellular organisms are practically immortal. Many of the infusoria may be dried up and apparently killed, but the addition of a drop of water revives these simple organisms, which will now swim about quite vigorously. When the complexity is great, as in higher animals, any local derangement puts the machinery out of gear; the penalty exacted for higher evolution is death.

Can we proceed even lower down in the scale of life, and reaching inorganic matter find in it some sign of irritability which had hitherto been regarded as the special characteristic of living matter? Has matter itself then the promise and potency of life? Otherwise how did life make its first appearance on earth? When the earth was a molten mass, there could have been no life as we know it now. It has been suggested that the seed of life was imported to this earth by the cosmic dust from other worlds; but this would merely transfer the difficulty backwards. Probably, at some critical period of the earth's history, the environmental conditions may have favoured the appearance of life out of non-life.

I was startled more than thirty years ago by certain quite unexpected characteristics of my detectors for electric waves. I found, for example, that under continuous stimulation by the oncoming message, the sensitiveness of the metallic detector disappeared. But on giving the receiver a sufficient period of rest, it became sensitive once more. On taking records of successive responses, I was surprised to find that they were very similar to those exhibiting fatigue in animal muscle. And just as animal tissue, after a period of rest, recovers its activity, so did the inorganic receiver recover after rest.

Thinking that a long rest would make my receiver more sensitive, I laid it aside for several days and was astonished to find that it had become inert. In fact, it had become lazy through lack of stimulation. A strong shock stirred it up again into readiness for response. Two opposite treatments are thus indicated for fatigue from overwork and for inertness from long passivity!

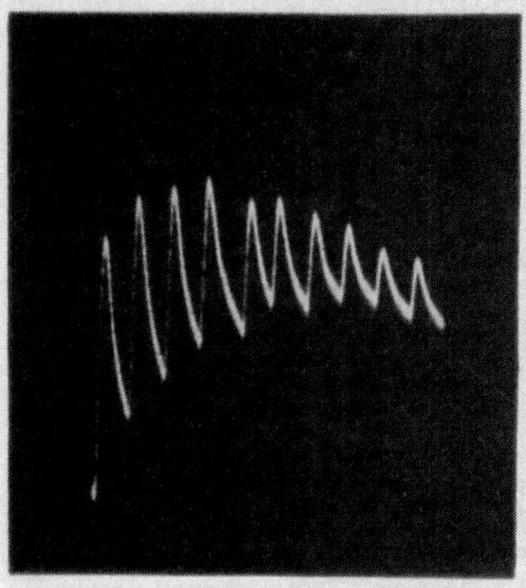

FIG. 36. Fatigue of Metals.

This was not all, for I further found that the application of certain stimulating drugs rendered my receiver extraordinarily sensitive so that it could record extremely feeble wireless messages, which it had failed to detect before. Other drugs depressed the sensitiveness, or abolished it altogether.

Having thus obtained some indications of reaction which reminded one of the responsiveness of living tissues, experiments were next carried out upon metallic wires, using the same method of electric response commonly applied to the investigation of animal tissues.

I now describe some of the more important results obtained with inorganic matter, metals for example. Metals, such as tin, copper, and even the chemically inactive platinum, were selected on account of their better conductivity, since this ensures larger electric response. As long as the piece of metal was at rest, there was no electric sign of excitation. But when it was excited in any way, a definite response was obtained. A feeble stimulation gave rise to a feeble response and the recovery from the state of excitation was found to take place in a short time. A stronger stimulation produced a larger response, the period of recovery being prolonged. A feeble stimulation, individually ineffective, became effective on repetition. All these results are strictly parallel to the characteristic responses of nerve and of muscle.

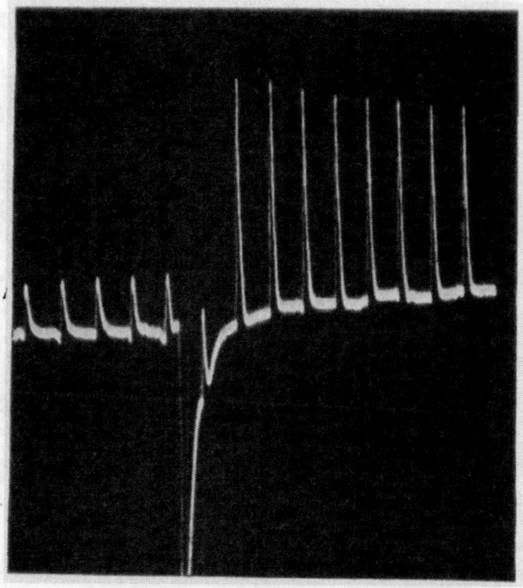

FIG. 37. Action of stimulant in enhancing the electric response of metal.

Uniform responses were obtained under uniform stimulation with suitable periods for complete recovery. But when the metal was subjected to long-continued stimulation, it showed fatigue (see fig. 36).

The effect of stimulants in enhancing the reaction will next be described. After taking the normal response of the metal to successive uniform stimulations, it is touched with the stimulating agent, a solution of sodium carbonate, with the result that there is an exaltation of the response (fig. 37). There are other stimulants which induce a still more astonishing increase of irritability.

Thus, as stated in this and in previous chapters, when animal, plant and metal have been subjected to the same questioning shocks they have in all cases given similar replies. They exhibit similar fatigue and show similar exaltation under stimulants. If further confirmation of the unity of reaction be needed, there remains one final test by which physiologists distinguish the characteristic phenomena of life. That which is living is capable of dying, and death may be hastened by poison. The pulses of response then wane until they cease altogether. Is it credible that we may in like manner 'kill' metals by the administration of poison?

Strange as it may appear, the electric response disappears altogether after application of poison. There remains the very curious phenomenon known not only in the investigation of physiological response, but also in medical practice, that of the opposite effects produced by the same drug when given in large or in small doses. Here, too, we have the same phenomenon reproduced in an

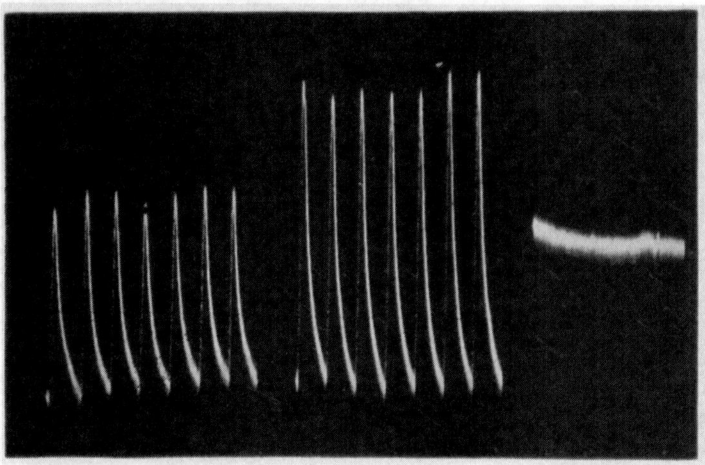

FIG. 38. Effect of poison on response of metal.
A minute dose enhances, and a large dose abolishes, the response.

astonishing manner even in inorganic response. A small dose of poison amplifies the response, while a large dose destroys it (fig. 38).

We have thus examined the autographic records of the living and non-living. How similar are the writings! So similar, indeed, that we cannot tell one from the other. We have watched the responsive pulses wax and wane, in the one as in the other. We have seen response sinking under fatigue, becoming exalted under stimulants, and being abolished by poisons, in the non-living as in the living.

Among such phenomena, how can we draw a line of demarcation and say, 'Here the physical process ends, and there the physiological one begins'? Such a line can hardly be drawn.

Do not the two sets of records of the living and non-living tell us of some property of matter common and persistent? Do they not show us that the same kind of molecular upset on stimulation occurs on both the inorganic and the living—that the physiological is closely connected with the physical—that there is no abrupt break, but a uniform march of law?

The dust particle and the earth, the plant and the animal, are all sensitive. Thus, with an enlarged cosmic sense, we may regard the million orbs that thread their path through space, as something akin to organisms, having a definite history of their past and an evolutionary progress for their future. We may then come to realise that they are by no means insensate clods, locked in the rigour of death, but active organisms 'whose breath, perchance, is luminous iron vapour, whose blood is liquid metal and whose food is a stream of meteorites.'

We shall henceforth turn with renewed courage to the investigation of mysteries which have long eluded us. For every step of science has been made by the inclusion of what seemed contradictory or capricious into a new and harmonious simplicity. Her advances have always been towards a clearer perception of underlying unity in apparent diversity.

It was when I came upon the mute witness of these self-made records and perceived in them one phase of an all-pervading unity that bears within it all things—the mote that quivers in ripples of light, the teeming life upon our earth, and the radiant suns that shine above us—it was then that I understood for the first time a little of that message proclaimed by my ancestors on the banks of the Ganges thirty centuries ago:

> They who see but one, in all the changing manifoldness of this universe, unto them belongs eternal Truth—unto none else, unto none else.

CHAPTER IX

THE CURVE OF LIFE AND DEATH

WHAT is the difference between life and death ? ' Tremulous with life,' we say. What do we mean by this ? In contrast rise the words ' As still as death.'

Life indeed is tremulous, because the living substance is constantly throbbing under the impact of those shocks from within which are vaguely called ' internal stimuli,' and shocks from without, or ' external stimuli.' But internal and external stimuli are in fact one, seen as it were from different points of view. All that is within the living organism has been acquired from outside. Into the primordial speck of life comes inpouring energy from the universe without. With the capital of a single cell, life begins its bank account, and this grows and waxes great as the accumulation of matter and energy increases. Multiplication and enlargement are the visible signs of acquired wealth. At full maturity, income and expenditure are balanced. After this comes loss greater than recuperation ; at the end bankruptcy and death.

An organism is actively living as long as it is responsive to the forces of its environment. It gives an answering twitch to the impinging stimulus. It does not necessarily give out all that it receives, for it may store a portion of the energy of the incident stimulus. And when, thus gathering by driblets, it has accumulated an excess of energy, this overflows or bubbles over in spontaneous movements. This process will be studied in greater detail in a subsequent chapter.

PLANT AS A MACHINE

From the physical aspect, an animal or plant may be likened to a heat-engine supplied with fuel. In order to find out how effectively the machine is working, we may attach to it an apparatus like my writing lever, which gives a graphic record of its activity. When the fuel runs short, the beat of the piston, and the up and down strokes of the record come to a stop. This stoppage from the exhaustion of the fuel is but temporary, for a fresh supply of fuel renews the movement. But the working of the engine may also be permanently stopped by some breakdown of machinery and interlocking of cogwheels. In the machinery of the living organism likewise, the activity of life may be temporarily stopped by the run-down of energy previously stored. The activity may also be permanently stopped by some internal clogging of the living machine.

Under the action of various poisons, the pulse-throb is gradually arrested in death. We may regard every fragment of the living substance as an aggregate of molecular machines. The change from a living to a dead condition is one of transformation from a state of molecular mobility to that of interlocked rigidity.

We see a man instinct with life suddenly falling into a death-like stupor under the action of a narcotic. By timely treatment he may be made to shake off the lethargy and return to life. But if there is delay, then life passes into non-life. The thinnest line thus demarcates life from death. What a crucial moment it is when the living particles are swinging in their unstable poise; a little tilt this way and the living machine is locked in the grip of death! In this supreme moment is played out the drama of life and its end. If we could, at this crisis, trace out the history of the molecular conflict, then and then only would be revealed to us the secret of death. That invisible internal conflict, the struggle between the forces of life and death, is sometimes vividly exhibited by the automatic record of the dying organism.

Visible Signal of Death

Discoloration is one of the signs of the death of the plant. A simple method of bringing about death is to place the plant in a bath and gradually raise the temperature, until at the fatal point it is scalded to death. If we take a portion of a flower, say the milk-white pistil of *Datura alba*, the normal white turns to brown at the occurrence of death. More striking still is the sudden discoloration of flowers of brilliant hue. The rich mauve petals of the Passion-flower are swiftly transformed, as if by magic, to pallid white. Thus if we immerse one half of the flower in a thermal bath, leaving the other half outside, the immersed half turns white at death, the rest retaining its purple hue. Similarly in the Indian crimson flower Sesbania, the red colour is transformed at death to a pale blue. The fatal temperature in normal specimens is more or less definite, being about 60° C. or 140° F. If the plant be subjected to fatigue by over-stimulation, the death-point becomes lowered, the degree of lowering depending on the extent of fatigue. Like ourselves, the plant when weary cannot struggle effectively against death. It is thus possible to map out, in a coloured petal, invisible zones of fatigue. For this we take a pair of stencil patterns in metal and place the highly coloured petal between the two. The two stencils are then connected with the two electrodes of an electric coil, and shocks are passed through definite areas of the petal, which thus becomes fatigued. On removing the stencil, we notice no visible change, though a latent image of fatigue is present at the region subjected to the electric shock.

If we now place the specimen in a bath at, say, 50° C. the fresh portion of the petal shows no change, for the death-discoloration takes place only at a temperature of 60° C. The fatigued area, however, dies at the lower temperature and exhibits local death by discoloration. Thus the invisible image of the pattern of fatigue is now seen visibly developed as the discolored pattern of death. This is the thermograph, the heat-record and map of death.

Is there anything resembling this mapping of incipient death in plants under natural conditions? Let us consider the variegations of certain leaves; the leaves of the Elephant Creeper of India show white patches in the midst of bright green. These patches represent, in reality, the pallor of approaching death, and if we keep these leaves under observation we find that it is at these particular spots that the tissue becomes disintegrated by decay which spreads outwards. The brilliant tint of leaves in autumn is but the hectic flush preceding death.

Electric Discharge at Death

I have already explained how changes which would have remained beyond our scrutiny can be detected by the electrical variations by which they are accompanied. Thus a sudden excitation gives rise to a negative electric change. I will presently show that an intense excitation is produced at the moment of death, with the result that there is an electric discharge at the moment of death of the tissue.

A simple method of demonstration is to take one half of a green pea, and connect its inner and outer surfaces with a galvanometer. The half pea is slowly raised in temperature in a heating bath. At the death-point of 60° C., an intense electric discharge passes through the organism. The electric change at death is very considerable, being often as high as 0·5 volt. If 500 pairs of half-peas are suitably arranged in series, the terminal electric pressure will be 500 volts, more than sufficient to cause electrocution of unsuspecting human victims. It is well that the cook does not know the danger she runs in preparing this particular dish, and it is fortunate for her that the peas are not arranged in series!

Memory Revival

We have next to consider a very interesting secondary effect in connection with the electrical spasm which we found immediately to precede death. I refer to that sudden revival of memory of which we have all heard. Here I can briefly refer to the enquiry which I carried out into the fundamental background of memory. After each shock of stimulus the responsive surface undergoes a molecular distortion, from which recovery takes place very gradually; this recovery is, however, never quite complete. Traces are left of the impression made by the stimulus and these remain as latent images, invisible even under microscopic examination. Under suitable conditions, however, this invisible script becomes conspicuous. I have thus been able to form impressions on a metallic plate of which no sign whatever can be detected, and yet when the place is subjected to diffuse stimulation, the invisible images are readily seen.

Similarly all the impressions made on our sensory surface remain dormant as latent memory images, which become revived under the impact of the internal stimulus of will. The revival of memory, then, is the result of a strong stimulation being thrown on the impressioned surface, so as to wake up the dormant images. Now we have seen that during the struggle of death an electrical spasm sweeps through every part of the organism and this strong and diffuse stimulation—now involuntary—may be expected to crowd into one brief flash a panoramic succession of all the memory images latent in the organism.

The Spasm of Death

The various symptoms of death in ordinary plants, such as drooping, withering, and discoloration, do not manifest themselves exactly at the moment of death, but at a later period. Even after a plant has been subjected to a temperature in excess of the fatal degree it may continue for a time to appear fresh and living. How, then, can it be possible to distinguish a living from a dead plant, and how to find the exact moment of transition?

Such discrimination is possible by watching for the disappearance of some reaction characteristic of the living condition. The ideally perfect method, however, would be the discovery of a reaction which, at the moment of death, underwent a sudden reversal to its opposite. There would then be not even that minor degree of uncertainty which is inseparable from the determination of the vanishing point of a waning effect. Such a perfect method I have been able to render practicable by the discovery of the death-spasm in plants analogous to the death-throe of the animal.

In securing the record of the onset of death in plants, two different methods have been found suitable. The first is to subject the plant to a continuous rise of temperature, until the fatal degree is reached. The second method, not so perfect as the first, is to apply a dose of dilute poison which proves fatal after a shorter or longer period, depending on the dose and the virulence of the poison.

The Death-Recorder

I have succeeded in devising a Death-Recorder by which the dying organism, by means of a continuous script, records its exact death-point. The recording apparatus is of an oscillating type; the smoked glass plate is made to oscillate to and fro by an electro-magnetic contrivance, thus producing a series of dots at every degree of rising temperature. For obtaining a death-record, we take a specimen of Mimosa (other plants may be employed with equal success) and subject it to a gradual rise of temperature in a water-bath. The expansive up-movement is represented in this record by a down-curve, while the contractile down-movement is shown by an up-curve.

Fig. 39. Life and death curve of Mimosa. Successive dots in down or expansive part of curve represent rise of temperature of 1° C. Spasmodic contraction causing inversion of curve takes place at 60° C.

The experiment was commenced at 25° C. The effect of increasing warmth of the bath on the plant, as on ourselves, is an expansive relaxation, and the curve is seen to move downwards. This continues until we note a short pause in the movement at or about 60° C. Then a sudden spasm—the death-spasm—occurs, the recording lever being jerked up with convulsive violence (fig. 39). Observe how sharp is the point of inversion. The entire record, then, is the curve of life and death. If the plant be cooled before the point of inversion, it survives; but once that point is passed, there can be no recovery. It was at the sharp turning-point that the conflict between life and death was decided.

In young and vigorous specimens this death-throe is very violent; with advancing age it is less so. In extreme old age the record of the curve of life passes almost imperceptibly into that of death.

If the plant has been fatigued by previous over-stimulation, the death-point is reached at a lower temperature. In a particular case the fatigued plant died at 37° C., the death-point being thus lowered by as much as 23° C.

Transmission of Death-Excitation

The plant may be regarded as a colony of living units, and it is possible to kill a portion of it without destroying the rest. What would happen if a portion of the plant were killed? If there is an intense excitation at death, then this should be transmitted and cause appropriate response of motile organs at a distance. This was demonstrated in the following experiment.

The lower end of a shoot of Mimosa was placed in the bath and the temperature raised. At the critical point, an intense excitatory impulse was generated which, travelling upwards, caused the fall of all the upper leaves in serial succession. That this excitation was due to the local death of the immersed portion was proved by the fact that the upper leaves re-erected themselves in the course of twenty minutes. Cooling the water of the bath and reheating it once more did not give rise to any more impulse, for the portion immersed was already dead. The plant was next lowered about an inch in the bath, and repetition of the experiment caused a new excitatory impulse, due to death of the fresh portion of the living stem; the response to transmitted excitation was exhibited once more by the serial fall of the leaves. This offers conclusive proof of an intense excitation which occurs in the tissue at death.

Similarly, a portion of tissue subjected to poison gives out an excitatory impulse at the moment of death. This occurs earlier under strong or virulent poisons. It is delayed under the action of a dilute poison, since this requires prolonged application to bring about the fatal result.

Continuity between Life and Death

Certain considerations seem to show that the phenomena of life and death are not entirely antithetic, but that there is a continuity which bridges over the chasm. After each shock, the organism becomes irresponsive or dazed for a time, after which it gradually recovers. The duration of insensibility is prolonged with the intensity of the shock. Sensibility and insensibility—tokens of life and death—thus alternate. Our life is, in reality, a succession of incipient deaths! From a moderate stimulation there is a quick restoration, but after an excessively intense shock there is no recovery. Death is an extreme case of stimulation.

In records of response of Mimosa taken under moderate, strong, and excessive stimulation, we find certain characteristic differences. In the first case, the recovery is completed in the course of fifteen minutes, as seen in the curve of recovery reaching the axis or base-line (see fig. 4). With a still stronger stimulus, the recovery is prolonged for an hour, the recovery-curve meeting the axis at a still greater distance. Up to this time, there is possibility of restoration. But the shock may be so strong as to prove fatal, and the violent spasmodic contraction proves to be the spasm of death. The line of .recovery is now parallel to the axis, and never meets it.

Let us examine the suggestive case of a real image of an object formed by a concave mirror. The reflected rays cross and meet at the axis, forming a real image. As the object is brought nearer and nearer, the image, like the line of recovery under stronger stimulation, is projected to a greater distance. A time comes when the reflected rays never meet and no image can be formed on this side of the surface of the mirror. Has it then completely disappeared? Not so, for the phantom is now transferred to the other side of the mirror! And after the supreme shock of death, though there is no restoration of life on this side of the Great Mirror by which we see Nature reflected, is it possible that there may be a restoration and renewal on the other side that is hidden from us?

These, our mute companions, silently growing beside our door, have now told us the tale of their life-tremulousness and their death-spasm in a script that we can read. May it not be said that their story has a pathos of its own beyond any that we have conceived?

In realising this unity of life, is our final sense of mystery deepened or lessened? Is our sense of wonder diminished when we realise in the infinite expanse of life that is silent and voiceless, the foreshadowing of more wonderful complexities? Is it not rather that science evokes in us a deeper sense of awe? Does not each of her new advances gain for us a step in that stairway of rock which all must climb who desire to look from the mountain-tops of the spirit upon the promised land of truth?

CHAPTER X

AUTOMATISM

It has been shown that the plant responds to even a feeble stimulus. Each such responsive movement in the plant, as in the animal, can be referred to a definite cause, being due to an antecedent stimulus. There are other movements which are more mysterious than this, inasmuch as they are produced seemingly without any cause. For instance, it would seem that the heart beats of its own accord. A sudden contraction of the heart is followed by an expansion, and this rhythm is maintained spontaneously throughout the duration of life. These automatic movements are referred to some unknown and mysterious internal cause. What, then, is the solution of the mystery of this automatism?

FIG. 40. Leaf of *Desmodium gyrans*.
The two small lateral leaflets exhibit spontaneous movements.

Although such automatic movements are usually associated only with animal life, yet similar activities are found even in plants. An example of this is found in the Telegraph-plant, *Desmodium gyrans*, which grows wild on the Gangetic plains. And surely no phenomenon could be more surprising than the ceaseless activity of the leaflets of this plant. Its compound leaf has three leaflets, a large terminal, and two small lateral ones (fig. 40). The small leaflets move up and down like the semaphore formerly employed for telegraphic signalling. On a warm day the leaflets dance continuously up and down, and the Indian peasant's belief is that they dance to the snapping of one's fingers. In reality, they require no such encouragement, for the minute pulvinus at the joint of the leaflet undergoes, of its own accord, a sudden contraction, followed by slow expansion and recovery.

I have been able to make a leaflet record its pulsating movements continuously for more than twenty-four hours at a time. The record for four hours is given in fig. 41. The extraordinary uniformity of these pulsations is indeed very remarkable.

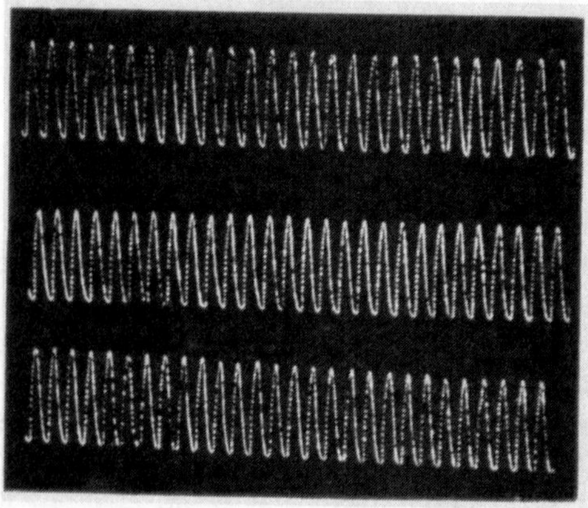

FIG. 41. Continuous record of pulsations of Desmodium leaflet for four hours.

AUTOMATIC PULSATIONS IN ANIMAL AND PLANT

There is an evident similarity between the automatic pulsation of the leaflet of the Telegraph-plant and that of

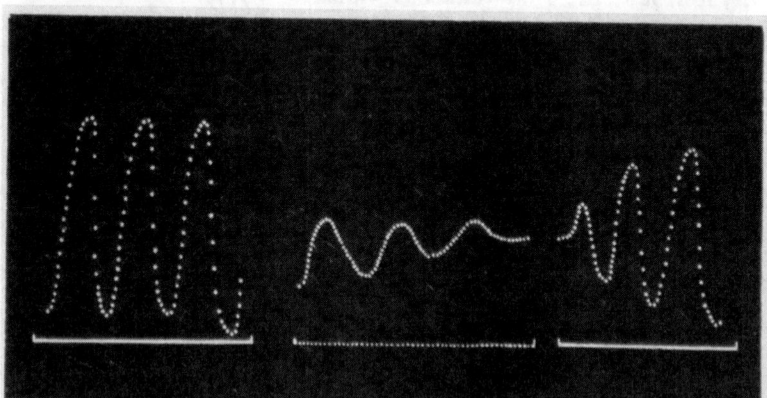

FIG. 42. Effects of irrigation and drought on pulsations of Desmodium.

The first series exhibits normal pulsation ; the second series, arrest under drought ; the third, revival of pulsation after irrigation.

the animal heart. Is this merely a superficial resemblance, or does it extend deeper? If it be a case of essential similarity, variation of external conditions should affect both alike.

The reactions of the cardiac tissue, which have been studied with the heart of the frog and of other animals, are such that a certain amount of internal pressure is necessary to start the pulsation. This is well shown in the renewal of pulsation in the quiescent heart of the snail on increasing the intracardiac pressure.

Similarly, the pulsations of the Desmodium leaflet become arrested under diminished internal hydrostatic pressure, consequent on the withdrawal of supply of water to the plant. Irrigation renews the pulsation (fig. 42).

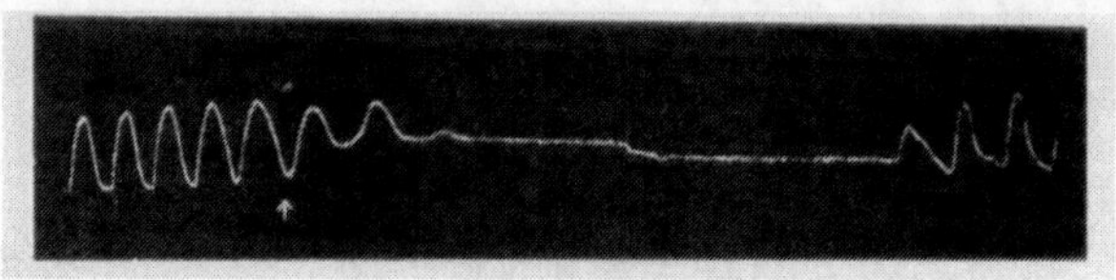

FIG. 43. Effect of vapour of ether.

Under anæsthetics, the heart-beat becomes arrested. The above record shows the stoppage of the pulsation of the Telegraph-plant under the action of ether-vapour. The pulsation was seen to come gradually to a stop. The ether-vapour was next quickly blown away and fresh air introduced into the plant-chamber; the plant revived after nearly a quarter of an hour and resumed its normal pulsation (fig. 43).

ANTAGONISTIC REACTIONS OF POISONS

There are numerous other parallelisms between the automatic pulsations of animal and plant tissues which will be given in a later chapter. I will here describe certain antagonistic reactions under different drugs. Poisonous acids, for example, arrest the pulsation of the heart, but this particular arrest takes place at diastolic expansion. Alkaline poisons also stop the beating of the heart, but in an opposite manner, that is to say, at systolic contraction.

That the action of these two poisons are antagonistic is further seen in the fact that when the heart-beat is arrested by one, it can be revived by the application of the other. This is a curious instance of one poison acting as the antidote of another.

It is wonderful to discover identical reactions in Desmodium. Poisonous acids arrest the pulsation at diastole, while alkaline poisons produce an arrest at systole. Finally, the arrest induced by either of these poisons can be counteracted by the antagonistic action of the other.

These experiments conclusively demonstrate the fundamental identity of the pulsatory mechanism in the animal and the plant. But the question still remains, what is the cause of these automatic movements?

Discovery of Connecting Link

We have hitherto been acquainted with a single responsive movement to a single stimulation; and now we find that there is quite a distinct class of phenomena in which movement takes place apparently without cause. Is there a hiatus between the two, or is there a connecting link, the discovery of which might possibly lead to an explanation of these mysterious automatic movements?

Such connecting links I have been able to discover in the plants *Biophytum sensitivum* and *Averrhoa Carambola*. Biophytum is a weed which grows in the neighbourhood of Calcutta. Its sensitive leaflets, arranged on the leaf-stalk in two long rows, show excitation by a twitching movement. Averrhoa is a large tree, the leaflets of which are also sensitive. They are arranged somewhat like Biophytum leaflets, on a long leaf-stalk in two rows. Ordinarily speaking, the leaflets are horizontally outspread. Each leaflet in both Biophytum and Averrhoa, when excited by a stimulus of any kind, responds by a fall, followed by recovery. The impulse caused by strong excitation of one leaflet may radiate to the neighbouring ones, and we are then presented with the most remarkable sight of a progressive rippling movement of all the sensitive leaflets.

MULTIPLE RESPONSE

Biophytum and Averrhoa exhibit characteristics of response which are precisely similar. To avoid repetition, I will describe in detail the reaction of Biophytum under increasing intensity of stimulus. I subjected a particular leaflet to stimulation by electric shock of increasing intensity, from $\frac{1}{20}$ of a unit to $\frac{1}{10}$, to $\frac{5}{10}$ and then to 1 and to 2 units. At $\frac{1}{20}$ of a unit the plant gave no answer, but when the stimulus was raised to $\frac{1}{10}$, a pronounced response was obtained. When the intensity was raised to 1 unit, the response was exactly the same. Hence the Biophytum leaflet exhibits response on the 'All or None' principle. It either responds to the utmost or not at all.

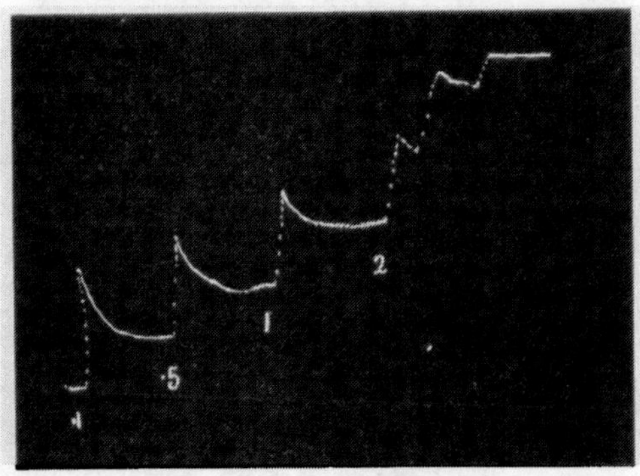

FIG. 44. Response of Biophytum leaflet to stimuli of ·1, ·5, 1, and 2.

Response is seen to be multiple with the last.

It is thus seen that 1 unit did not produce any larger response than that produced by a stimulus one-tenth of that intensity. What became of the excess of energy that impinged on it in the form of stimulus? It is not necessary to suppose that in every instance the whole energy of the stimulus absorbed is employed in causing the movement; some of it may be wasted as heat. But on the other hand, it is conceivable that the excess may be stored up for the time being, to find subsequent expression. To take a physical illustration, the energy stored up in a compressed spring may on release give rise to persistent oscillations. The question arises then, whether there may be an analogous storage of energy in the leaflet of Biophytum after excessive stimulation. Supposing this to take place, the superfluous energy may be utilised in the form of storage not immediately discerned by an observer; the stored energy may, however, find expression later in the form of rhythmic or repeated movements. That this is actually the case is seen in the response to a stimulus of intensity of 2 units. Here we find that a single strong stimulus has given rise, not to a single but to multiple responses in recurrent series (fig. 44).

These multiple responses are given under any form of strong stimulus, such as a mechanical blow, an electric shock, the shock of heat, strong rays of light, or chemical stimulation. The persistence of the multiple activity, moreover, depends on the energy of the shock that impinges on the

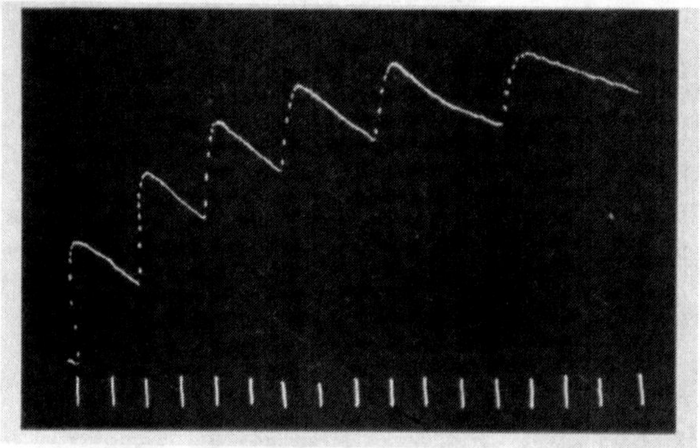

FIG. 45. Multiple response in Averrhoa under a single strong electrical shock.

plant, a moderate shock giving rise to a few, and a strong or long continued shock giving rise to many recurring pulsations (fig. 45). The response thus echoes, as it were, or reverberates. There is here an analogy to the vibrations given by a tuning fork; when gently struck, it vibrates and emits sound for a short time, but when strongly excited by a stronger blow, it gives a more persistent note.

SEMI-AUTOMATISM

Under natural conditions, the plant is exposed to the action of various stimuli, supplied by its environment. It is exposed to heat, to the action of light, to the mechanical stimulus of air-currents, and to the action of various chemical agents present in it or absorbed by it. From the joint action of these external sources of stimulation, the energy stored up by the plant becomes sufficiently great to cause an excitatory overflow. It was our want of sufficiently penetrative analysis of the previous history of the plant that led to the assumption that these movements were self-caused. The internal stimuli for such automatic movements are, in reality, external stimuli which have become trapped.

We may, for convenience, divide plants into two classes, ordinary and automatic. Biophytum, under normal conditions, comes under the first category, since it gives a single response to a single moderate stimulation. But

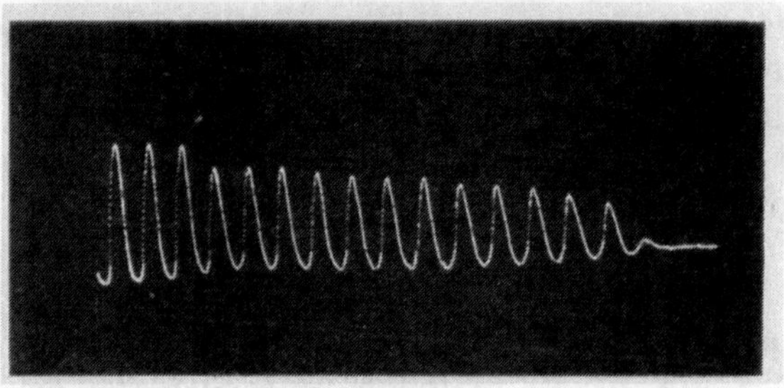

FIG. 46. Gradual arrest of pulsation of Desmodium after depletion of stored energy.

under strong stimulation it gives multiple responses, which are apparently automatic. There is thus a continuity between ordinarily and automatically responding plants.

Let us now see whether a pronouncedly automatic plant can be brought down to the condition of an ordinary plant, and if it is the store of excess energy that enables Desmodium to exhibit spontaneous movement. I took a Desmodium plant and shut it off from the stimulating energies of its environment. Depletion of its store of energy soon brought pulsation to a standstill (fig. 46). A fresh accession of energy was now found to renew the pulsation; a feeble stimulus produced a single response; a stronger stimulus gave rise to a series of multiple responses. When restored to the normal stimulation of the environment it regained its automatic activity.

It is the stored energy, then, that bubbles over in

apparent spontaneity. This fact casts a flood of light on various activities until now deemed irrational. A healthy baby, on waking after its meal, throws out its limbs over and over again in a rhythmic manner, expressive of overflowing energy. This is also seen in children of a larger growth, when an intensely pleasurable stimulation will cause them to dance for joy, a form *par excellence* of rhythmic activity. Again, one has often been struck by the long series of zigzag flourishes, characteristic of certain signatures. This multiple rhythmic response is not improbably the expression of an overflowing self-esteem !

Rhythmic Sensation

This multiple response is not merely characteristic of the grosser mechanical movements, but is equally true in the subtler realm of sensation.

Let us think for a moment of the effect on the retina of strong stimulation by light. We stare for a moment at the incandescent filament of a bulb and then close the eyes. Even after the cessation of the stimulus, the after-effect of the light persists as a series of strong visual impressions, the bright filament appearing and disappearing in alternation for a considerable length of time. This means that the intense stimulus of light has caused multiple responsive sensations.

The same is true in like manner of all forms of mental stimulation. These, when very intense, are apt to repeat themselves and become persistent. And it is in vain that we seek to escape from their recurrence. We are dogged by our own thoughts, and even in our dreams they return to us. Like other forms of living matter, nervous tissue is tuned or primed to added responsiveness by the very stimulus that impinges upon it. It is as the cumulative effect of repeated stimulations that the activity of nervous matter ultimately becomes automatic, as it is seen to be in many processes from the birth of thought to inspiration.

Transient and Persistent Spontaneity

The persistence of pulsatory activity on the cessation of external stimulation depends on the capacity for storage.

In this respect, I find that plants exhibit two distinct types. In one, the overflow is initiated with very little storage; and here, the unusual display of multiple activity soon comes to a stop. To maintain such a type in the rhythmic condition, constant stimulation from outside is necessary. Plants of this type are extremely dependent on outside influences, and when such sources of stimulation are removed, they speedily come to an inglorious stop. Biophytum is an example of this kind. In the second type of automatic plant-activity, a long continued storage is required before an overflow can begin. But in this case the spontaneous outburst is persistent and of long duration, even when the plant is deprived of any immediately exciting cause. These, therefore, are not so obviously dependent as the others on the sunshine of the world. The Desmodium plant furnishes an example of this type.

Have we not here a suggestive parallel to a certain phenomenon known among literary and artistic people as inspiration? For the attainment of this exalted condition, it is also necessary to have a previous storage for subsequent effervescent overflow. Inspiration is, after all, an example of fullness and overflow of life without any apparent effort. If this be so, aspirants to this condition might well decide in whose footsteps they will choose to tread—those of Biophytum, with its immediate dependence on favourable outside influences and ephemeral activity, or of those of Desmodium, with its characteristic of patient and long-enduring accumulation of forces, which afterwards find uninterrupted and persistent expression.

A continuity of response has thus been traced, both in the plant and animal, from simple, through multiple, to autonomous response. A feeble or moderate stimulation evokes a single response, whether this be mechanical or psychical. Strong stimulation, on the other hand, gives rise to multiple response in various forms. Excess of energy derived from the absorption of various environmental stimuli, bubbles over, as it were, and shows itself in spontaneous rhythmic response. This is variously manifested, now in the pulsation of the Desmodium leaflets, again in the beating of the heart, and elsewhere in echoing sensation and thought.

A living organism may, from one point of view, be

regarded as simply a machine. But in order to keep this living machine at work, in all those complex and wonderful ways of which it is capable—from mechanical movement, through throbbing sensation to spontaneous thought—something more than mechanical perfection is necessary. It can only throb with life when placed in constant communion with all the forces of the universe about it.

CHAPTER XI

THE RECORD OF GROWTH

THE phenomenon of growth furnishes us with another example of automatic activity. Can any similarity be discovered between this and other modes of spontaneous activity? Do variations of external conditions modify all automatic activities in a similar manner?

The question of growth and its modifications under external change is a matter of great practical importance, for the world's supply of food depends on the growth of plants. It is, therefore, of the highest importance to be able to discover all those conditions which are favourable to growth. In what way do external agencies modify it?

MEASUREMENT OF GROWTH

The essential difficulty of the investigation arises from the extraordinary slowness of growth. Even the proverbial slow-moving snail moves two thousand times faster than the tip of a growing shoot. The average rate of growth is about $\frac{1}{100000}$ inch per second, a length which is half that of a single wave of sodium light!

Even with the magnifying growth-recorders hitherto employed it takes a very long time to detect and measure its rate. For accurate investigation of the effect of any given agent on growth, it is necessary to keep all other variable conditions, such as light and warmth, strictly constant during the whole period of the experiment. We can keep these conditions absolutely constant for only a few minutes at a time. Experiments which require several hours for their completion are, therefore, subject to serious errors which vitiate the results.

The only satisfactory method is one that reduces the period of the experiment to a few minutes; that, however, necessitates the devising of an apparatus for a very high magnification, and for the automatic record of the magnified rate of growth.

The High Magnification Crescograph

I have succeeded in devising an apparatus by which growth becomes instantly visualised. This produces a

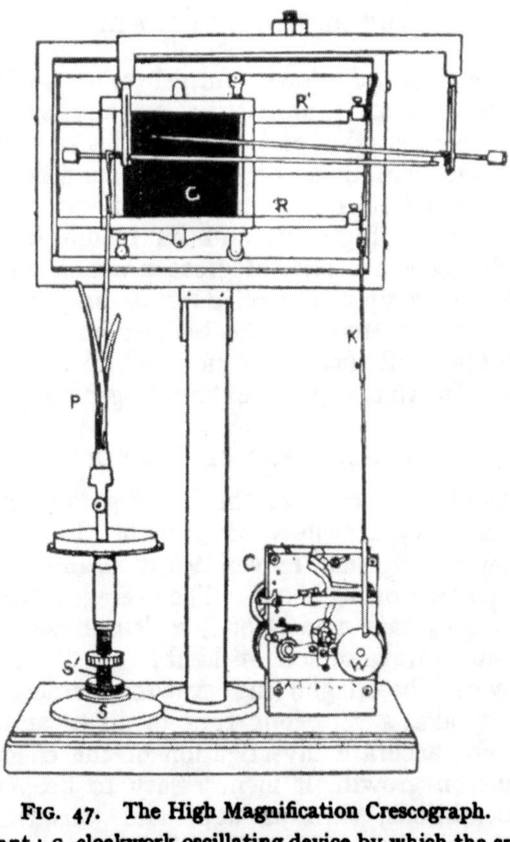

Fig. 47. The High Magnification Crescograph.
P, plant ; C, clockwork oscillating device by which the smoked glass plate is made to move to and fro.

magnification of ten thousand times, which would increase the rate of a snail's pace to that of an express train.

The High Magnification Crescograph consists of a system of two levers. The first magnifies a hundred times, and the second enlarges the first a hundredfold, the total magnification being thus 10,000 times (fig. 47). The record is taken on a smoked glass plate kept moving to and fro by an oscillating device, at regular intervals of time varying from one to ten seconds. The recording plate is kept moving laterally at a uniform rate. A curve of growth is thus obtained. A stimulating agent which enhances the rate of growth, produces a flexure of the curve upwards; a depressing agent, on the other hand, lessens the slope of the curve.

The method of high magnification has revealed many interesting characteristics of growth hitherto unsuspected. It has also established the essential similarity of growth to other modes of automatic activity.

PULSATIONS OF GROWTH

Like other modes of automatic activity, growth is found to be rhythmic or pulsatory. This is seen in the

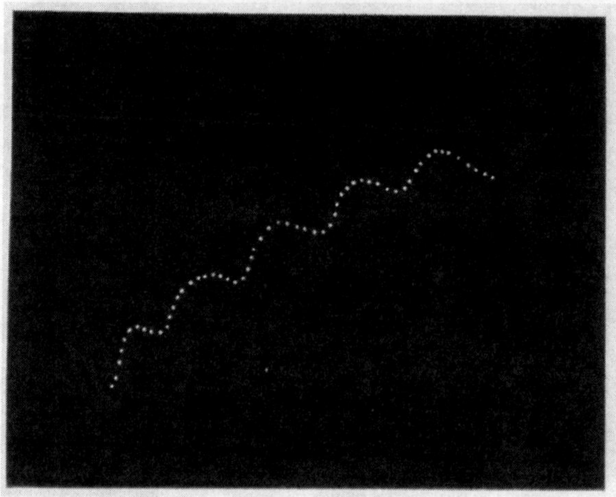

FIG. 48. Record of Pulsation of Growth taken with a High Magnification Crescograph.

record given in fig. 48, which shows a sudden uplift followed by a slower recoil, the amount of recession being about one-fourth of what had been gained. The difference between the two represents the permanent growth. The growth process is, therefore, not steady, but like the wavelets of a rising tide.

Effect of Internal Hydrostatic Pressure

In the Desmodium leaflet the pulsating activity is arrested when the internal pressure is reduced, as under conditions of drought; the pulsations are, however, revived after irrigation.

Similarly, growth, under conditions of drought, is brought to a standstill. It is interesting to find that the application of a few drops of water to the root now gives rise to a few pulses of growth, which, however, soon come to an end. More profuse irrigation ensures continuous growth. I have already stated that increased internal pressure, an enhanced state of turgor, accelerates growth. Turgor of the tissue is maintained by the pumping activity which supplies sap to the growing region. It may be stated in a general way, that any condition which increases the rate of sap-ascent also increases the turgor and the rate of growth. Conversely, conditions which retard the rate of ascent or cause a diminution of turgor also inhibit growth.

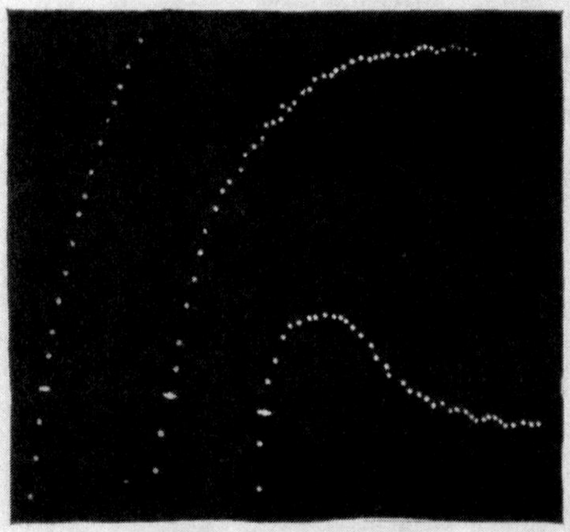

Fig. 49. Effect of electric stimulation of increasing intensities of ·5, 1, and 3 units on growth elongation.
The last record exhibits the actual shortening of the growing organ under stimulation.

Effect of Stimulation

I explained in a previous chapter that stimulation produces in a pulvinated organ a diminution of turgor and contraction (see p. 34). The effect on a growing organ is essentially similar. I reproduce records of growth under increasing intensity of stimulation. Application of electric stimulus of 0·5 unit produced a slight down-flexure in the curve, indicating a moderate retardation of growth.

An intensity of 1 unit induced a greater retardation, whereas a stimulus of 3 units induced an actual contraction, as shown by the reversal of the curve (fig. 49). It will be seen that stimulation induces a reaction which is opposite to that of normal growth-elongation. This may conveniently be described as 'incipient' contraction, for under increased intensity of stimulus it culminates in an actual contraction.

A striking result discovered in the course of this investigation is that while a stimulus of moderate intensity retards growth, a feeble stimulus accelerates it. Two diametrically opposite effects are thus produced under strong and feeble stimulation. We shall find a wider application of this principle in the action of drugs which may be regarded as chemical stimulants.

Similarity of Response of Pulvinated and Growing Organs

I have shown that under a sufficiently strong stimulus, growth-elongation is reversed into an actual contraction. After a period of rest, growth is renewed. The record of response and recovery shown in fig. 50 is essentially similar to the response and recovery of a pulvinated organ. The difference is merely one of degree.

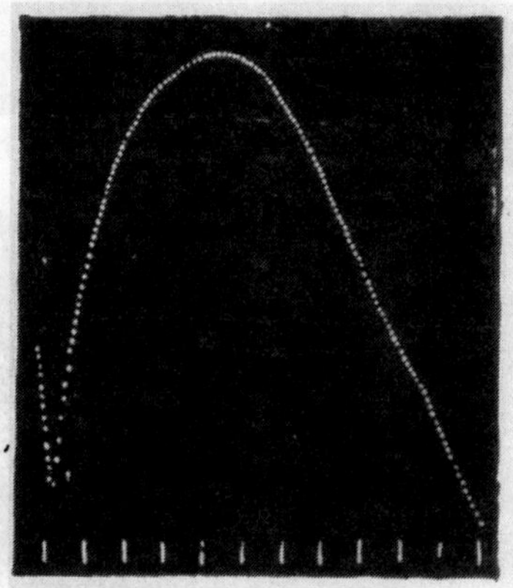

FIG. 50. Contractile response of growing organ under electric shock.

Successive dots at intervals of 4″. Vertical lines below indicate intervals of one minute (magnification 1,000 times).

There is a prevalent wrong assumption that different modes of stimulation induce dissimilar reactions. The results of my investigations prove that this is not the case. All modes of effective stimulation of a certain intensity, by contact, by electric shock, or by light, produce the same responsive contraction.

The growing organ subjected to diffuse stimulation undergoes, as we have seen, an equal contraction all round, on account of which it exhibits a shortening. But if one side of the organ be subjected to local stimulation, then that side alone will contract, producing a bending of the organ to one side. This phenomenon will be described in greater detail in a later chapter.

EFFECT OF VARIATION OF TEMPERATURE

Change of temperature has a marked effect on the rate of growth. It is depressed by a fall and becomes arrested at a critical low temperature. Conversely, warmth produces an astonishingly rapid increase of growth up to an optimum temperature, beyond which it becomes once more retarded; at 60° C. a death-spasm occurs, after which there is permanent abolition of growth.

EFFECT OF ANÆSTHETICS

A small dose of a mild anæsthetic like ether stimulates growth. The first part of the curve (fig. 51) indicates by

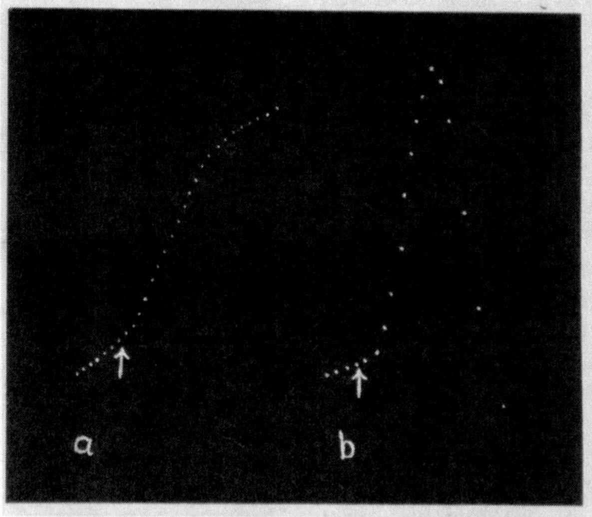

FIG. 51. Effect of Anæsthetics on Growth.
(a) Enhancement under small dose of ether; (b) preliminary enhancement followed by spasmodic death-contraction under the action of chloroform.

its slope the normal rate of growth; diluted vapour of ether produces a sudden erection of the curve, indicating an enhancement of the rate of growth. Under application of chloroform, the quantity of vapour absorbed at the beginning being slight, the effect of a small dose is seen to be an acceleration of growth. As more chloroform is absorbed, the result is a retardation instead of acceleration, leading ultimately to a spasmodic contraction which causes a reversal. The apex of the curve demarcates life from death: after this reversal, spots of discoloration appear in the plant; these spread very rapidly and the specimen becomes wilted as a consequence of death.

The experiment on the action of ether shows the possibility of reviving dormant growth by a mild stimulant.

Effect of Chemical Stimulants

Any marked advance in scientific agriculture is only possible if we succeed in discovering agents which greatly enhance the activity of growth. Only a few stimulating agents have been used for this purpose, whereas there are numerous others of the action of which we have been profoundly ignorant. The rule-of-thumb method hitherto employed in the application of a few chemical stimulants and of electricity has, moreover, not been uniformly successful. The cause of this anomaly is found in the discovery of an important factor, namely, that of the amount of the dose, which has hitherto not been taken into account. Thus we find that while a particular intensity of electrical current accelerates growth, any excess above a critical point retards it. The same is true of chemical stimulants. A striking result was obtained with certain poisons, which in normal doses kill the plant, but in excessively minute doses acted as an extraordinarily efficient stimulant in promoting growth. In fact, plants behave in this respect like human beings. Investigations on plants thus open out fresh lines of research in Pharmacology and Medicine. The High Magnification Crescograph, moreover, enables speedy testing of the action of manurial agents, which takes only a few minutes instead of a whole season, thus avoiding the errors arising from changing conditions of protracted experimentation.

Not only is growth modified by the intensity of the

stimulation, physical or chemical, but also by the point of its application and by the tonic condition of the plant.

POINT OF APPLICATION OF STIMULUS

The effect of stimulation is modified not only by the intensity but also by its point of application. Growth and all modifications associated with it take place in a more or less restricted region, a little behind the tip of the growing organ. When a stimulus is applied *directly* at this growing region, the result is a retardation of growth, which under strong stimulation culminates in actual contraction

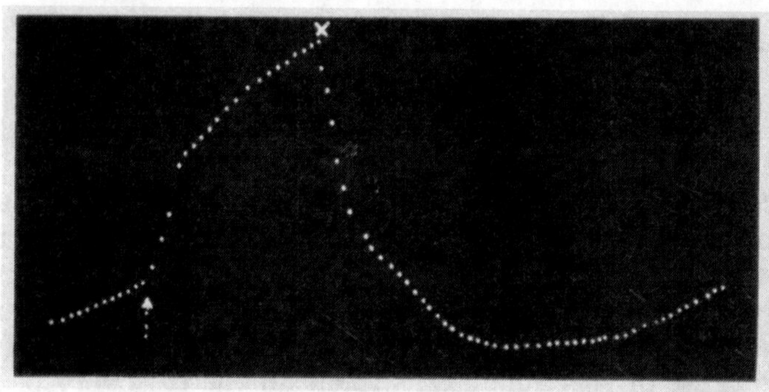

FIG. 52. Effects of indirect and direct stimulation on growth. Indirect stimulation at dotted arrow enhances the rate. Direct stimulation at cross induces a reversed response by contraction.

and shortening of the organ. But a totally unexpected result occurs when the stimulus is applied at a distance from the responding region of growth. Under this indirect stimulation, the rate of growth becomes actually enhanced. This is illustrated by the records given in fig. 52. The normal growth is represented by the upward slope of the first part of the curve. Indirect stimulation applied at the point marked by the arrow is seen to produce a sudden erection of the curve, proving the acceleration of the rate of growth. Stimulation was next applied directly at the point marked by the cross, with the result that the growth was not only arrested, but the organ exhibited a marked contraction, shown by the reversed curve (see also fig. 62).

The Law of the Effects of Direct and Indirect Stimulation is:

Direct Stimulation induces Contraction: Indirect Stimulation induces the opposite effect of Expansion.

MODIFYING EFFECT OF TONIC CONDITION

In medical practice the anomaly is frequently met with, that the same drug induces diametrically opposite effects in different individuals. The cause of the anomaly lies in the fact that the tonic level or constitution of different individuals is not the same. For rational treatment, it is essential to take into account the constitution of the patient, for the reaction of a given drug is greatly modified by his tonic condition.

In experimenting with plants, I artificially modified the tone or constitution of different batches of similar seedlings. One was kept normal for reference, another depressed to a sub-tonic condition, and the third raised to an optimum state of exceptional vigour. A measured dose of dilute poison was applied to the three batches of plants. The normal plants survived after a period of struggle. The weaker specimens succumbed at once without any struggle. But the reaction of the vigorous specimens was quite different; the toxic agent not only failed in its illegitimate work, but actually exalted the growth of its intended victims!

It will be noted how numerous are the factors which modify growth. It is only by studying the separate effects of these individual factors that we can unravel the complications which at first sight appear hopeless. The reduction in the duration of experiments, rendered possible by a ten thousandfold magnification of growth-movement, has enabled us to discover characteristics of growth hitherto unsuspected. I hoped to find other subtle reactions indicated by the minutest variations in the rate of growth; but such effects could only be discovered by increasing still further the sensitiveness of the method of detection.

CHAPTER XII

THE MAGNETIC CRESCOGRAPH

THE method described in the last chapter showed how changes produced in the rate of growth can be detected by the resulting flexures in the recorded curve. But it is conceivable that minute variations in the rate of growth may be attended by changes in the slope of the curve too slight to be detected by that method.

THE BALANCED CRESCOGRAPH

It was therefore necessary to devise a new method which would instantly show, by the up or down movement of an indicator, the accelerating or retarding effect of an agent on growth. I was successful in carrying out this idea by devising the Method of Balance. The plant was made to descend at the exact rate at which its growing tip was rising. A regulating device had to be introduced, analogous to the compensating movement of an astronomical telescope, which neutralises the effect of the earth's movement round her axis once in twenty-four hours. The problem was, however, more difficult; for instead of compensating a definite rate, adjustments had to be made for balancing widely varying rates of growth in different plants and even in the same plant under different conditions.

In the Balanced Crescograph, a train of revolving clock-wheels, actuated by the fall of a weight, lowers the plant exactly at the same rate at which it is growing. By the gradual turning of a screw S to the right or to the left, the rate of compensating fall can be continuously increased or decreased. The rate of growth is thus accurately compensated, so that the growing tip of the plant remains exactly at the same level (fig. 53). The tip is attached, in the usual manner, to the High Magnification Crescograph, and the Recorder now dots a horizontal line instead of an ascending curve as in the previous method. The apparatus thus balanced is rendered extraordinarily sensitive. The minutest change induced in the rate of growth by the environment is at once indicated by the upset of the balance shown by the up and down movement of the curve. The

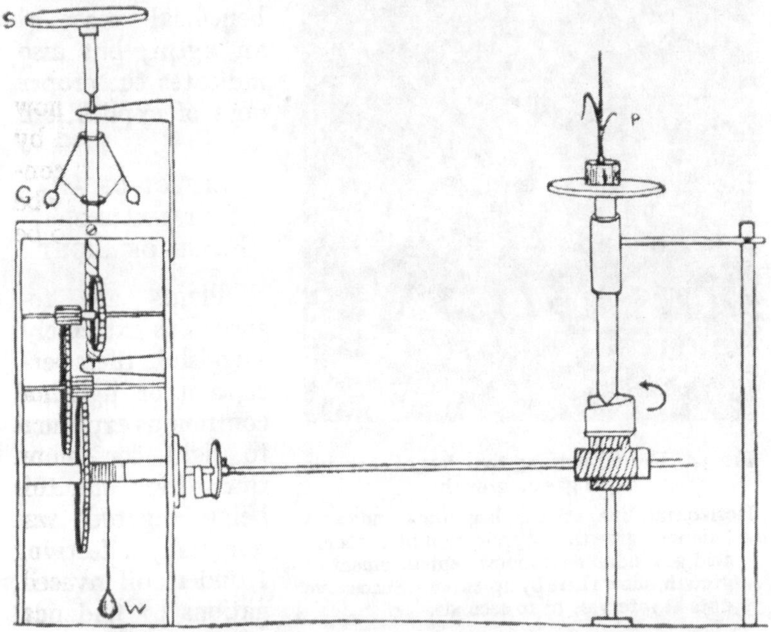

FIG. 53. The Balanced Crescograph.

Compensation of growth-movement produced by equal subsidence of the holder containing the plant (P). Adjusting screw (S) regulates the speed of the governor (G). W, heavy weight actuating clock-work.

method is so extremely sensitive that it detects and records variations in the rate of growth so excessively minute as $\frac{1}{1000}$ millionth of an inch per second.

The delicacy of this method is shown in the record of the effect of carbonic acid gas on growth (fig. 54). A jar filled with this gas was emptied over the plant; the gas, on account of its weight, fell in a stream and surrounded the plant. The record shows that this gave rise to an immediate acceleration of growth which continued for 2½ minutes; under the continued action of the gas, the preliminary acceleration was followed by retardation of growth, as shown by the down curve. The Balanced Crescograph not only exhibits the beneficial effect of an agent, but also indicates the proper dose of application.

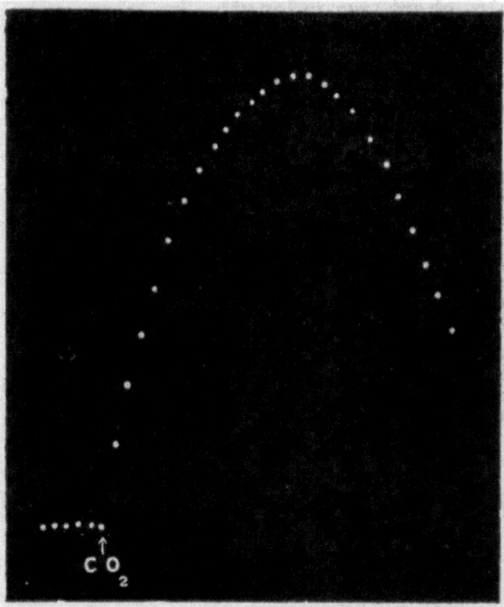

FIG 54. Record showing the effect of carbonic acid gas on growth.

Horizontal line at the beginning indicates balanced growth. Application of carbonic acid gas induces temporary enhancement of growth, shown here by up-curve. Successive dots at intervals of 10 seconds.

Effect of an Instantaneous Flash of Light

Plants are regarded as extremely sluggish in their perception of light, a continuous exposure to light for more than five minutes being regarded as minimally effective. I undertook investigations to find out whether the plant responded to light-exposure of excessively short duration. It is impossible to conceive of anything more fleeting than a single flash of lightning. I subjected the growing plant balanced on the Crescograph to an artificial flash of lightning, that is to say, to the light emitted by a single electrical spark between two metallic balls. The plant perceived and responded to this light of incredibly short duration, as was manifest from the upset of the balance and the resulting automatic script made by the plant.

Wireless Waves and Growth

The plant has thus been shown to perceive not only a feeble stimulation, but also stimulation of excessively short duration. The next subject of investigation was the range of perception of the plant when exposed to the various visible and invisible rays. The plant responds to ultra-violet light, with its extremely short wave-length, by slowing the rate of growth. The retarding effect of light on growth declines towards the less refrangible rays, the yellow and red. As we proceed further into the infra-red region,

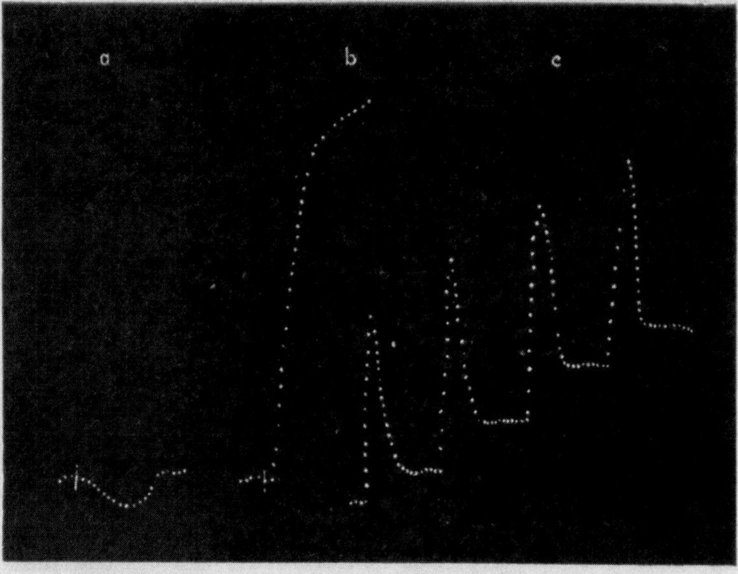

FIG. 55. Record of response of plant to wireless stimulation.
(a) Response to feeble stimulation by acceleration of growth; (b) response to strong stimulation by retardation of growth; (c) response to medium stimulation—retardation followed by recovery. Down-curve represents acceleration, and up-curve retardation of growth.

we come across the vast range of electric radiation, the wave-lengths of which vary from the shortest wave that I have been able to produce (0·6 cm.) to others which may be miles in length. There thus arises the very interesting question whether plants perceive and respond to the long æther-waves, including those employed in signalling through space.

The Balanced Crescograph was employed to record the possible change induced in growth by the action of wireless waves. The results of my investigations show that wireless waves produced characteristic variations in growth, depending on the intensity of stimulus. Feeble waves produced an *acceleration* in the rate of growth. Very strong waves produced a retardation, the effect persisting for a long time after the cessation of stimulation. But with waves of medium intensity the induced retardation was followed by a quick recovery (fig. 55). The perceptive range of the plant is inconceivably greater than ours; it not only perceives but also responds to the different rays of the vast ætherial spectrum.

Perhaps it is as well that our senses are limited in their range. For life would otherwise be intolerable under the constant irritation of these ceaseless waves of space-signalling to which brick walls are quite transparent. Hermetically-sealed metal chambers would then have afforded us the only protection.

Magnetic Amplification

The High Magnification Crescograph magnifying 10,000 times, in conjunction with the Method of Balance, gave results beyond all expectation. This should have satisfied one, but man is never satisfied, and it is his insatiable longing to surpass his previous achievements that is the real impetus for all progress. I therefore set about inventing a super-crescograph; if by using two levers I had succeeded in raising the magnifying power from a hundred to ten thousand times, a third lever might raise it proportionately higher. But the attempt failed completely. The weight and therefore friction at the points of contact which linked the different lever-systems were too greatly increased, and the theoretical advantage was wholly nullified in practice. I was, therefore, forced to work out a new solution. Material contact having proved unworkable, I had to discover a new method of linking which would be immaterial and therefore frictionless. This was found in a magnetic contrivance, in which a delicately poised system is upset by the movement of a magnetic lever in its neighbourhood.

This will be understood from fig. 56. SN is a light magnetic rod supported on a fulcrum. The short arm of the lever is attached to the tip of a growing plant represented by W. Growth-elongation lowers the N end of the magnetised rod, which causes increasing deflection of the suspended magnetised needle *ns* with the small attached mirror. The sensitiveness is greatly increased by the

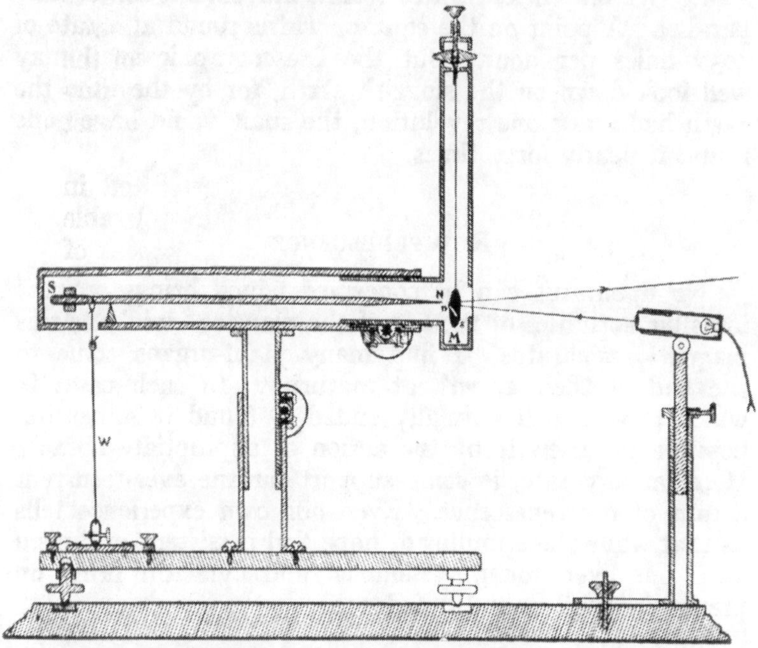

FIG. 56. Diagrammatic representation of Magnetic Crescograph. S–N, magnetic rod supported on fulcrum; short arm of lever attached to growing plant W. Growth-elongation lowers the N end, which causes increasing deflection of the suspended needle with attached mirror M. Deflection magnified by reflected spot of light.

employment of a perfectly astatic system of needles for the single *ns*. A line of light reflected from the mirror is thrown on to a distant scale. The travelling of the reflected light shows the growth-movement magnified from ten to a hundred million times.

It is difficult for the mind to grasp magnification so stupendous. Some concrete idea of it can, however, be obtained by what the speed of the proverbial snail will become when magnified ten million times by the Magnetic

Crescograph. For this enhanced speed there is no parallel even in modern gunnery. The 15-inch cannon of the *Queen Elizabeth* throws out a shell with a muzzle velocity of 2360 feet per second. But the Crescographic snail would move at a speed 24 times faster than the cannon shot. We may next turn to cosmic movements for a closer parallel. A point on the equator whirls round at a rate of 1037 miles per hour. But the Crescographic snail may well look down on the sluggish earth, for by the time the earth had made one revolution, the snail would have gone round it nearly forty times.

Rejuvenescence

No apparatus can be conceived which brings out the invisible activities of the life of the plant so vividly as this magnetic apparatus. I find many plant-organs come to the end of their growth at maturity. In such cases in which growth had normally ended, I found it sometimes possible to renew it by the action of appropriate stimuli. Here, at any rate, is some support for the ever-recurrent dream of rejuvenescence. Even our own experience tells us that while the stimulus of hope and persistent optimism keep one ever young, pessimism and cynicism bring on premature senility and decadence.

Life in a State of Suspense

The Magnetic Crescograph is remarkably suitable for the demonstration of growth before a large audience. No experimental conditions for exhibition of growth could have been more difficult than those prevailing in the depth of an English winter, when the plants were in a state of hibernation. In spite of this, they were made to shake off their lethargy, and the rate of resulting growth was exhibited by the indicating line of light rushing across a 10-foot scale in the course of some twelve seconds, the actual rate being less than a hundred thousandth part of an inch per second.

Other experiments, even more striking, on the action of drugs in depressing or enhancing the rate of growth can be easily exhibited. The growth of the plant under normal conditions is seen magnified as a line of light rushing across the scale. By a depressing agent the growth becomes paralysed and the line of light is brought to a stop, but a dose of a stimulant instantly removes the depression. The life of the plant thus becomes subservient to the will of the experimenter; he can exalt or depress its activity; he may thus bring it near the point of death by application of poison, and when the plant is hovering in an unstable poise between life and death, resuscitate it by the timely application of an antidote.

It is true that man is very imperfectly equipped to make daring adventures on the great seas of the Unknown. Of all possible notes of sound, only eleven octaves are audible to him, and a single octave of light circumscribes his vision. Even then the size of the ripple of visible light imposes an impassable barrier. He will never be able to see objects smaller than a fifty-thousandth part of an inch, which is the length of a single light-wave. These limitations have not deterred him, but have, on the contrary, spurred him on to greater efforts in his explorations in the region of the invisible. The mysterious movements of life are not to remain to him inscrutable for all time; but his untiring effort and single-minded pursuit will reveal to him the secret that lies behind all manifestations of life.

CHAPTER XIII

WOUNDED PLANTS

As we step out into the garden there are spread before us multitudinous plants, silent and seemingly impassive. Various shocks impinge on them and they may suffer wounds and death. In what way is the plant affected when the shock has been very excessive, as after a severe wound ?

I undertook three separate investigations into the effect of wounds. The object of the first was to ascertain the effect of injury on growth ; of the second, to demonstrate the reaction of wounds on the throbbing pulsation of the Telegraph-plant ; of the third, to analyse the paralysing effect of wounds.

Effect of Wound on Growth

Using the High Magnification Crescograph I subjected a plant to various irritations, from a rough touch to severe wounds, and observed the resulting changes in the rate of growth. In a particular experiment, I first took the record of normal growth, and then irritated the plant by rubbing it with a rough piece of cardboard. Growth was now retarded to two-thirds of its normal rate. The plant was next given a rest of fifteen minutes to recover from the irritation, and the restoration of the rate of growth was found to be only partial. It was a full hour before recovery was complete. Rough touch retards growth, and the rougher the treatment the longer the plant takes to recover from it.

Of this I may give an instance which puzzled me for a long time. I had been attaching numbers of plants to my Crescograph to obtain their records of growth. Though the apparatus was in perfect adjustment, yet no growth could be recorded. A plant happened to be left attached to the apparatus over night, and I found, to my surprise, that the plant which had failed on the previous day to show any growth, now exhibited it vigorously. I then realised that the unavoidable rough handling, in tying the plant to the apparatus, was enough to produce an immediate inhibition of growth, which was renewed only after recovery from the irritation. The procedure now adopted was to mount the plant as gently as possible, and to give it two hours' rest before taking the record. With this precaution, there was no difficulty in obtaining satisfactory results.

Pin-Pricks and Knife-Wounds

I next set out to find the effect of pin-pricks. These caused a greater irritation than friction or rough handling. Growth was now retarded to nearly a quarter of the normal rate, and the recovery took a proportionately longer time. Even after a full hour, growth recovered to only two-thirds of the normal rate.

A more severe wound was inflicted by a longitudinal slit made by a knife; this reduced the growth to about one-fifth of the normal rate. The wound-effect of a transverse cut was far more severe. Such a cut inhibited growth for a very long time. In sensitive specimens it actually produced a convulsive contraction.

Effect of Thrashing on Backward Growth

Corporal punishment is evidently not conducive to growth, a fact which might be seriously pondered over by schoolmasters! Truth, however, compels me to confess that it may have its uses. For my investigations show that while in actively growing specimens shocks cause a retardation of growth, in others, where the rate of growth is *below par*, stimulation revives and accelerates the rate. This fact may probably explain the custom among certain Indian peasants of giving a good thrashing to the young crop when it is lagging in growth!

For reasons at present obscure, certain plants remain stunted or backward in growth, the branches and leaves presenting an unhealthy look. Lopping off the offending limb is then found to be good for the plant. A severe shock renews growth that had become arrested.

Effect of Wound on Pulsating Leaflets

The leaflets of the Telegraph-plant, as has been already stated, exhibit automatic pulsation. When the small leaf-stalk carrying the leaflets was detached from the parent plant and the cut end of it placed in water, their pulsation was arrested by the shock of the operation. The shock-effect of the wound gradually disappeared, and the pulse-throb revived and continued for nearly twenty-four hours. Death, however, found an unguarded spot at the wound, and its march, though slow, was sure. The death-change ultimately reached the throbbing organ, which became permanently stilled. Experiments have been undertaken to arrest this march of death. The problem is intimately connected with the proper understanding of the conditions which lie behind life, and the other conditions under which the molecular cogwheels become arrested in the rigor of death. The experiments already carried out have been so far successful as to prolong, under proper treatment, the throbbing pulsations of the plant for more than a week.

Motor Paralysis

A severe wound paralyses the motor function of the pulvinus of Mimosa. On cutting off a short piece of stem bearing a leaf, the shock-effect was transmitted to every part of the parent plant, and all the leaves fell down and remained depressed for a considerable length of time. The detached piece with its cut end placed in a nutrient solution was also depressed, as shown by the fall of its leaf. The subsequent history of the parent plant and of the detached piece exhibited a significant difference.

The parent plant, paralysed by the wound, recovered slowly. A record was taken of the response of one of its leaves. The first record in fig. 57 (1) is the normal response of the leaf before the infliction of the wound. The leaf fell down, after the severe wound, to the position marked in the record by a cross. The paralysing effect of the wound was proved by means of testing shocks, the response being in each case inscribed by the automatic recorder.

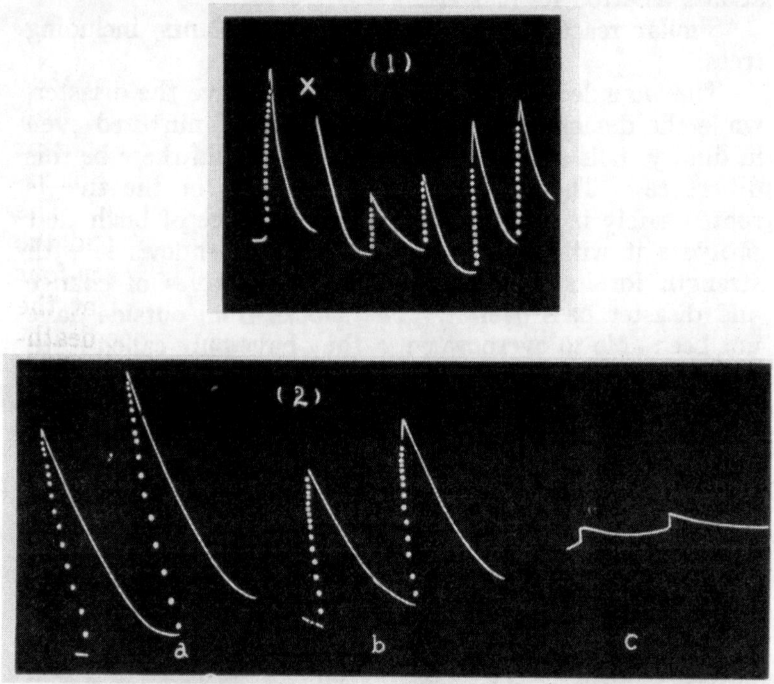

FIG. 57. The effect of wound (1) on parent plant and (2) on detached shoot.

(1) First response normal. The × shows contractile fall of the leaf after wound. The subsequent responses exhibit slow but perfect recovery.
(2) (*a*) Vigorous response of leaf of the detached shoot 4 hours after section ; (*b*) depression after 24 hours ; and (*c*) marked depression and final abolition of response at death after 48 hours.

The excitability remained depressed for nearly two hours, after which the plant gradually recovered its normal excitability in a 'staircase' manner. The normal functions of the plant afterwards became fully restored.

The leaf of the detached shoot, fed with the nourishing solution, soon held itself up with an attitude almost of defiance! In its newly found freedom from the parent plant it was unusually energetic in its responses. This vehemence lasted for a whole day, after which a curious change crept in; the vigour of its responses began rapidly to decline. The leaf, hitherto erect, fell over, death having at last asserted its mastery.

Similar reactions take place in other plants, including trees.

The wounded plant is thus able to survive the disaster, while the detached shoot with its free leaf, nurtured even in luxury, falls a prey to death. Why should there be this difference? The reason is that the plant or the tree is rooted safely in its own soil. It is the place of birth that provides it with proper nourishment and endows it with strength for its struggle in life. Many waves of change and disaster pass over it. The shocks from outside have not been able to overpower it; they have only called forth its nascent powers. It has met external change by counter-change. The decaying and the effete, such as worn-out leaves, have been cast off as changing times called forth its powers of readjustment.

The tree also derives an additional power from its racial memory. Every particle of the inconspicuous embryo within the seed thus bears the deep impress of the mighty banyan tree; and so the sprouting seedling forces its roots into the yielding earth to anchor more safely; the stem rises high against the sky in search of light, and the branches with their canopy of leaves spread out in all directions.

What is the strength, then, that has conferred on the tree its power of endurance and enabled it to emerge victorious from the struggle of life? It is the strength derived from the place of its birth, its power of perception and quick readjustment to change, and its inherited memories of the past.

CHAPTER XIV

THE SENSE OF DIRECTION IN PLANTS

THE movements induced by external stimulation by contact and by light have been described in the last chapter. There is little vitality in the outside epidermis or skin of the plant-body upon which the stimulus impinges; hence arises the necessity for special contrivances by which stimulation at the outside may reach the deeper and more active layers of cells in an intensified form. There are thus various adaptations by which mechanical stimulation or the stimulus of light may induce irritation in the living cells. When we pass the hand gently over our skin, this causes little stimulation; but if there is a thorn in the flesh, the leverage exerted by it will produce an intense irritation. We have tactile hairs or bristles for accentuating the external stimulus of contact. In plants likewise, there are tactile hairs and bristles. In Mimosa, such hairs occur on the under-side of the pulvinus, and by their lever-action cause irritation of the motor tissue. In many of the tendrils, again, there are tactile pits by which the stimulus of contact is accentuated. As regards the perception of light, Haberlandt has shown that in many leaves the epidermal cells are lens-shaped, so that the incident light becomes focussed on the sensitive protoplasmic layer and thus facilitates the perception of light.

The stimuli enumerated above possess energy and deliver, as it were, a blow on the plant; we can easily make out the direction of stimulation in these cases. But when we come to the effect of the stimulus of gravity, we encounter many difficulties. It should be remembered that gravity only stimulates a plant-member, when that member has been removed from its normal relation to the vertical. The difficulties which confront us are, firstly, that we cannot see the force of gravity as we do rays of light; secondly, the force of gravity acts only indirectly through a special sense-organ, as will be subsequently explained.

The problem before us is, first, the determination of the means by which geotropic irritation is effected, and second, the exact localisation of the sense-organ by which the plant is able to perceive when it is out of the vertical, thereby directing its movement so as to become once more erect.

Let us begin with consideration of some of the conditions which produce either an acceleration or a retardation of the reaction.

Effect of Temperature

High temperature is detrimental to geotropic reaction. I became aware of this from the records obtained under otherwise parallel conditions in Calcutta during summer (temperature 98° F.) and in the hill-station of Darjiling. The geotropic action was far more pronounced in the cooler climate of the hills.

The enhanced geotropic reaction at a moderately low temperature and the depressed reaction at a higher temperature, offer a satisfactory explanation of the movement of the Praying Palm (see fig. 32). It will be remembered that the inclined stem became increasingly erected by its enhanced geotropic reaction during the fall of temperature, while rise of temperature caused the opposite reaction of declension.

Effect of Ether

The geotropic reaction can, moreover, be extraordinarily exalted by the application of moderate doses of ether vapour. I took two similar leaf-stalks of *Tropaeolum majus* (Nasturtium) and subjected them to geotropic stimulation for an hour, N being normal and the other, E, etherised. Figure 58 shows the enhanced reaction under ether as compared with the normal specimen. While the normal one had turned up to a moderate extent, the other had completed not only 90° erection, but had gone beyond it.

In figure 58A the record shows that the normal rate of geotropic response is greatly enhanced after application of ether as evidenced by the sudden erection of the curve.

We now return to the puzzling question as to the nature of the sense-organ by which plants perceive the vertical

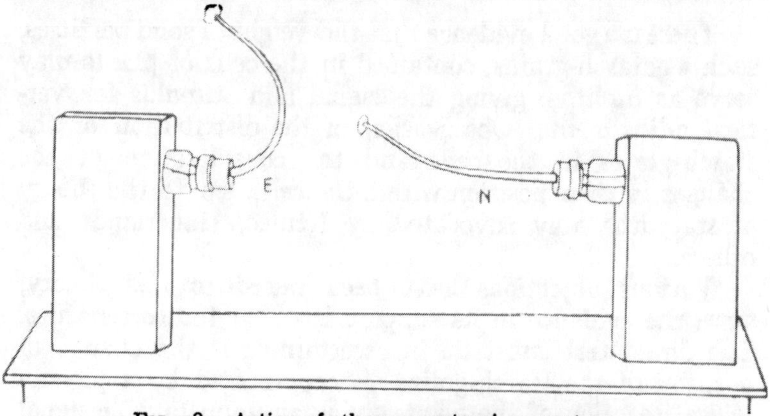

FIG. 58. Action of ether on geotropic curvature.

direction and thus obtain an indication of how to readjust themselves. Masons get their idea of the exact direction of the force of gravity by the use of the plumb-line, that is to say, by a hanging weight. We ourselves obtain the sense of direction by means of the semi-circular canals in our internal ear. The contained fluid exerts varying pressure in different positions, and thus gives us an idea of the vertical direction. In the lower animals, the lobster, for example, there are 'otoliths' and sand-grains mingled with tactile hairs, which by their vertical pressure give these animals their sense of direction in space.

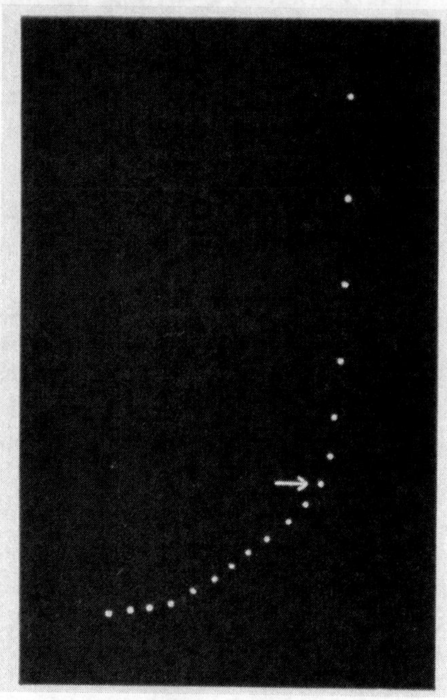

FIG. 58A. Effect of ether in enhancement of geotropic response.

Statolith-Theory

There is a good evidence that the weight of solid particles, such as starch-grains, contained in the cells of plants may serve as otoliths, giving the signal and stimulus for vertical adjustment. Observation of the distribution of the starch-grains in the cells and the consideration of the changes in their position within the cells, led to the theory of statoliths ably advocated by Nêmec, Haberlandt and others.

Various objections have been raised to this theory, since the evidence in its support is of an indirect nature. The direct test must lie in ascertaining if the change of position of the starch-grains is accompanied by a physiological reaction of the plant, giving an unmistakable signal of perception of geotropic stimulation by the plant as it is displaced from its normal vertical position. It is necessary, in fact, to make an exploration inside the still living plant to discover if the particular layer containing the statoliths is the one most irritated by the geotropic stimulation.

The Electric Probe

I have been able to solve this problem by the device of the Electric Probe. Suppose G and G′ (fig. 59) to be the layers of cells concerned in a stem with the perception of the stimulus of gravity, G and G′ being the longitudinal section of a hollow circular cylinder. The Electric Probe consists of an exceedingly fine platinum wire, enclosed in a capillary glass tube, the Probe being electrically insulated except at the extreme tip. When the Probe, suitably connected with a galvanometer, is slowly thrust into the stem, the galvanometer, by its deflection, shows the state of irritation, if any, of every layer of cells through which it passes. When the stem is vertical, the exploring Probe, during its transverse passage through the stem, detects no sign of local irritation.

The result is different when the stem is displaced from the vertical to a horizontal position. The geotropically sensitive layer G now perceives the stimulus and becomes the focus of excitation as evidenced by the negative electric response registered by the galvanometer. The excitation at the perceptive layer irradiates into the neighbouring cells in radial directions with an intensity diminishing with distance. The intensity of responsive electric change is observed to decline in both directions, outwards and inwards, from the focus.

The distribution of the excitatory changes initiated at this perceptive layer, and irradiated in radial directions, is represented by the depth of shading, the darkest shadow being on the perceptive layer itself (fig. 55, right fig.). Had excitation been attended in the plant with a change of

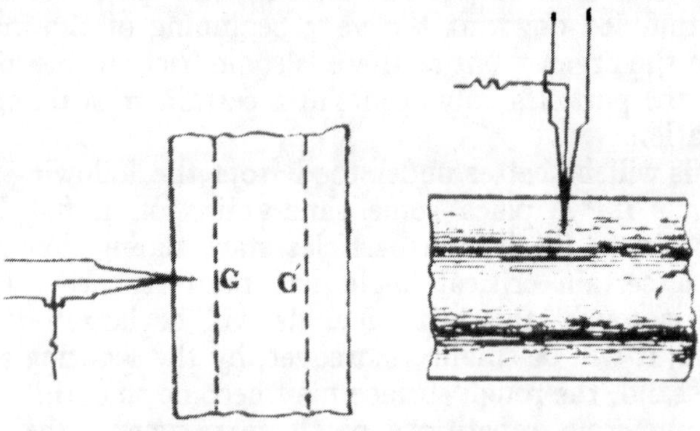

FIG. 59. Localisation of the geo-perceptive layer by means of the Electric Probe. Diagram represents the geo-perceptive layer in unexcited vertical and in excited horizontal position.

light into shade, we should have witnessed the spectacle of a deep shadow spreading from the perceptive layer over the different layers of cells during displacement of the organ from vertical to horizontal; the shadow would have been observed to disappear on the restoration of the organ to the vertical position.

After localising the perceptive layer by the electric probe, I made a section of the stem, and found that the cells of this layer contained large-sized starch-grains which by their weight were instrumental in causing stimulation. Thus the statolith-theory received independent confirmation.

Sliding of Particles at Critical Inclination

I have been able to devise another crucial experiment for testing of the statolithic theory. The inclination of the stem to the vertical was gradually increased until an abrupt electric response to geotropic stimulation occurred at a critical angle of inclination. This will be understood from the following consideration. In a vertical position, the particles are at the base of each cell; excitation can only occur when they fall abruptly on to one side of the cell. Had there been no viscous friction, the particles would have tumbled down at the very beginning of the inclination of the organ; but as there is some friction, the abrupt fall of the particles only occurs at a certain critical angle of inclination.

This will be better understood from the following illustration. If we place some sand-grains on a flat board and begin to tilt it, the particles start sliding down only after a certain critical angle has been reached. If the board is rough, this critical angle will be large; if it is smooth, it will be small. Moreover, by the scouring action of the sand, the rough surface may become smoothed down after numerous repetitions of the experiment, the result being a diminished degree of the critical angle.

This is precisely what takes place in plants as regards their critical angle for the immediate perception of geotropic stimulation. Experiments which I carried out with various plants show this angle to be about 31°; there is no electrical response to geotropic stimulation until the organ is tilted to this angle. But when the experiment is repeated with the same specimen, the critical angle is reduced to about 25°.

The results of these experiments is, therefore, to strengthen the evidence in favour of the statolith-theory.

CHAPTER XV

PLANT-MOVEMENTS IN NATURE

THE plant is constantly adjusting itself, by appropriate movements, to the various forces of the environment which are acting upon it. Such movements are very slow and the final results become marked only after a lapse of time. When taking a walk in the garden we are struck by the constant changes that are taking place in the posture of various organs of plants. Here is a tendril that has touched the twig of a neighbouring shrub (fig. 60); the contact has been perceived by the tendril, which has clasped the support and twined round it, thereby helping the plant to climb up so as to get as much light as possible. There, a potted plant in a dark shed, which has so bent and twisted its leaves that their upper surfaces are fully exposed to the incident light coming through a hole in the shutter (fig. 61). In yet another place, the plant has turned away so as to avoid excess of light. Here a plant has been accidentally upset and the horizontally laid stem is bending upwards so as to grow erect once more.

These movements, induced by the stimuli of the environment, are extremely varied and complicated. The forces in operation are manifold:—the influence of changing temperature, the stimulus of contact, of gravity, and of light, visible and invisible. Again, an identical stimulus produces sometimes one effect and at other times precisely the opposite. It thus appears hopeless to unify these very diverse phenomena, and there has been a tendency towards a belief that there is not any general physiological reaction, but that it is the individuality of the plant that determines the choice of the movement for its own advantage. Such a teleological assumption is no explanation at all.

Various descriptive phrases have been in use to

designate the diverse movements, such as positive heliotropism, when the organ turns towards the light, negative heliotropism when it moves away from it. Similarly, the terms positive and negative geotropism have been employed to denote the movement of different organs towards or away from the earth. These merely descriptive phrases must not, however, be regarded as offering any real explanation of the phenomena. Such misuse of descriptive phrases 'may even be mischievous,' protests Bayliss in his book on physiology, 'leading to the belief that new knowledge has been obtained when a phenomenon is described by a name derived from the classical tongue instead of in English.'

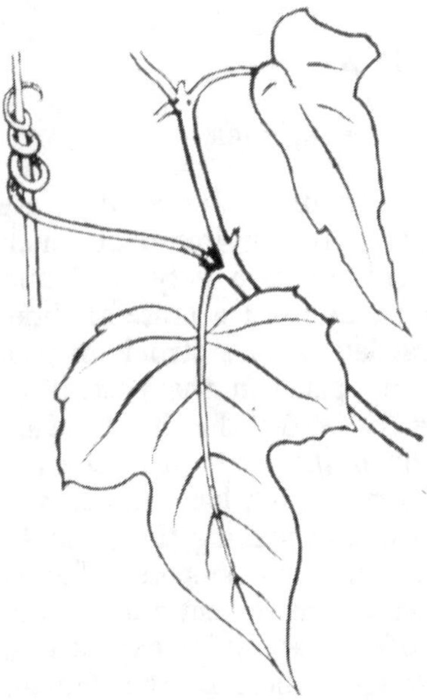

FIG. 60. The twining of a tendril.

It may perhaps be possible to offer a rational explanation of the varied movements of plants. The baffling character of the problem is, in reality, due to the combined action of different factors, the individual reactions to which are unknown to us.

In this connection there are two questions to be investigated, namely, the reactions under different kinds of stimulation, and the curvature induced by one-sided stimulation.

Much confusion has arisen from the uncertainty as to the effects of different modes of stimulation; the stimulus of contact, for example, may produce a specific reaction different from that of light. Such specific reactions have, in fact, been assumed for different stimulations. In reality, there is no such difference, for results described in a previous chapter show that all kinds of direct stimulation of a certain intensity give rise to contraction, actual or incipient; for instance, this incipient contraction is expressed in a growing

FIG. 61. Leaves of Tropaeolum (Nasturtium) facing the light.

organ by retardation of the rate of growth. I have already explained how there are various contrivances by which external stimulation reaches the sensitive tissue in the interior, often with increased intensity (see p. 103).

Positive Curvature

I shall next consider the question of the bending or curvature of an organ induced by one-sided stimulation. Let us take a concrete example and imagine the right side of a stem to be acted upon by a stimulus, say, mechanical contact or light. The right side of the stem is stimulated *directly* and consequently there is induced a local contraction, actual or incipient. The normal rate of growth of the right side is now retarded, so that if the left side grows as usual, the stem will become curved, the stimulated side being concave. The stem will thus become bent towards the stimulus, a bending which is often described as *positive* tropism. I shall presently show that there is a second contributory factor which accentuates the curvature.

In the case of positive tropism described above, we centred our attention on the right side of the organ which was *directly* stimulated. What happens on the left or distal side which was not directly stimulated? Does it remain passive, or is some *indirect* effect transmitted to it? This indirect effect, if it exists at all, will either help or hinder the production of curvature due to the direct action of stimulus. Nothing has hitherto been known about the effect of indirect stimulation.

Effect of Indirect Stimulation

With regard to the effect of indirect stimulation on normal growth in length, I have shown (fig. 52) that while direct stimulation induces a diminution of turgor, with contraction and retardation of growth, indirect stimulation produces the diametrically opposite reaction of increase of turgor, with expansion and enhancement of rate of growth. This is diagrammatically represented in fig. 62 *a*, *b*. Direct stimulation induces contraction and shortening of the organ, while indirect stimulation causes expansion and elongation.

Parallel effects may be expected when a stimulus acts on one side of the organ. It is the change of turgor induced by stimulation which is the basis of all responsive movements. I have succeeded in devising a striking experiment demonstrating the opposite reactions induced by direct and indirect stimulation.

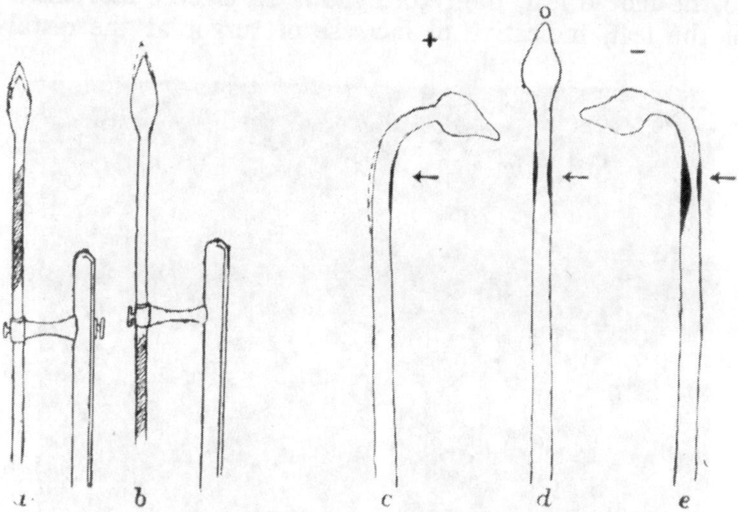

FIG. 62. Effects of Direct and Indirect Stimulation.

(*a*) Stimulus applied directly at the growing region induces retardation of growth or contraction, as represented by dotted line. Stimulated area represented in this and following by shading.

(*b*) Stimulus applied indirectly (at some distance from growing region) gives rise to acceleration of growth and expansion.

(*c*) Stimulus applied to right side of organ causes contraction of that side, thus giving rise to positive curvature towards stimulus.

(*d*) Excitation transmitted to the opposite side causes neutralisation.

(*e*) Excitation caused by intense stimulation is transmitted across and thus reverses the positive curvature to negative, *i.e.* away from stimulus.

We shall now consider the opposite effects of direct and indirect stimulations. In Mimosa, diminution of turgor is visually manifested by the fall of the leaf, increase of turgor by a movement of erection. Figure 63 gives a diagrammatic view of the stem of Mimosa with an attached leaf, at the base of which is the motile pulvinus. If the stimulus, say strong light, be applied on the right-hand side, the pulvinus is directly stimulated and the leaf undergoes a fall. If then the stimulus be transferred to the opposite side of the stem at a point diametrically opposite to the motile leaf, the record shows an erectile movement of the leaf, indicative of increase of turgor at the distal

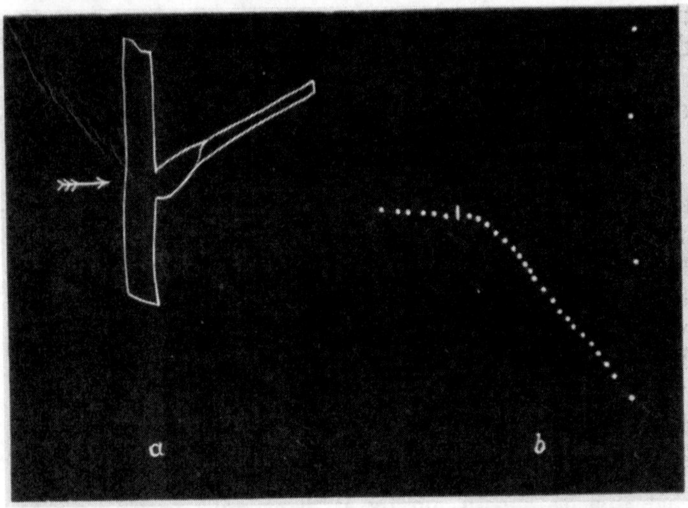

FIG. 63. Effect of one-sided Indirect Stimulation of stem of Mimosa.

a Diagrammatic representation of experiment.
b Record showing preliminary erectile response due to indirect stimulation, and subsequent excitatory fall caused by transverse conduction of excitation after continuous stimulation.

side. The effects of direct and indirect stimulation are thus opposite. The effect of continuous indirect stimulation in causing the fall of the leaf will be presently explained.

Thus one-sided stimulation by light causes a contraction of the proximal and expansion of the distal side. The two effects conspire to produce a positive curvature towards the stimulus (fig. 62 *c*). This normal bending towards the light is described as positive heliotropism.

Similar effects are produced under other modes of stimulation. Thus when a tendril is stimulated on one side by contact, then the side touched grows more slowly, and the opposite side faster, than the normal. The effects on the two opposite sides thus conspire to bring about a movement towards the stimulus and the curling of the tendril round the support.

It is, therefore, not necessary to assume diverse sensibilities and reactions to different modes of stimulation. All stimuli only produce contraction when they are direct and sufficiently intense. The simple law that direct stimulation induces contraction and retardation of growth, and indirect stimulation expansion and acceleration, explains all the 'positive' tropisms or turning of the organ towards the stimulus.

When an organ turns away from light, the phenomenon is called 'negative heliotropism,' the assumption being that the organ is possessed of a specific sensitiveness of a quite different kind. But an identical organ cannot be possessed of both positive and negative sensitiveness. And yet, if an organ is continuously exposed on one side to strong light, it is found that it moves at first towards the light, then the over-excited organ begins to turn away, as if to avoid an excess of light. The same organ is thus positively heliotropic for a time and negatively heliotropic afterwards.

Bending away from Stimulus

The organ bends *towards* a light of certain intensity, it bends *away* from light of greater intensity. This latter result is due to the percolation of excitation from the proximal to the distal side of the organ.

In regard to the transmission of excitation, the ordinary tissue of a plant is not such a good conductor of excitation as a nerve: but the difference is one of degree only, not one of kind. Nervous tissue transmits excitation to a distance, even under feeble stimulation, while conduction takes place in ordinary tissues with difficulty and under strong and continued stimulation. Thus the effect of moderate stimulation remains more or less localised at the directly stimulated side of the stem. But the excitation caused by excessive or long continued stimulation may overcome the block, and thus percolate and traverse the organ in a transverse direction.

The effect of this transversely transmitted excitation is demonstrated by the record of response of the leaf of Mimosa (see fig. 63), in which a point of the stem directly opposite to the responding leaf was subjected to continuous stimulation by light. The first effect was the erectile response of indirect stimulation; excitation was then conducted across the stem and caused the sudden fall of the leaf.

In the case of the bending of the stem under one-sided action of light, the first effect is a positive curvature. But under continued stimulation, the transversely conducted excitation induces contraction of the further side, causing neutralisation of the former bending (fig. 62 *d*).

Sometimes the near or proximal side becomes fatigued by over-excitation, while the far or distal side is still fresh. The result of this is not merely neutralisation, but an actual bending away from the light by the greater contraction of the distal side (fig. 62 *e*).

In an actual experiment carried out with a seedling of Indian corn (*Zea Mais*), the plant, under continuous action of strong light from one side, exhibited at first a movement towards light. This reached its maximum after fifty minutes. The plant then began to turn back and return to the position from which it started. This neutralisation was completed in a further period of forty-three minutes. After this, the response of the plant was a negative one.

The movements of life, then, are not capricious, but are determined by very definite physiological reactions. The elusiveness of the problem which had baffled us was due to ignorance of the numerous factors that are in operation. It has been shown how seemingly opposite results are brought about by the mere difference of the intensity and duration of stimulus. In actual conditions of life, numerous other complicating factors are also present. Thus the sign of the response is modified by the point of application of stimulus, by the transverse conductivity of the organ, and by the tonic condition of the plant. It is the different combinations of these factors that produce the numerous variations in the resulting response, which at first sight appear so perplexing.

The movements described in this chapter are such as are induced by one-sided stimulation ; these, however, by no means exhaust all the phenomena of plant-movement, for there are others which are induced, not by one-sided, but by diffuse stimulation exercised by the environment. A striking example of this is found in the periodic opening and closing of the flower of Water-Lily, fully described in the next chapter.

CHAPTER XVI

THE NIGHT-WATCH OF NYMPHÆA

THE poets have forestalled the men of science. Why does the Water-Lily keep awake all night and close her petals during the day? Because, say they, the Water-Lily is the lover of the Moon, and as the human soul expands at the touch of the beloved, so the Lily opens out her heart at the touch of the moonbeam and keeps watch all night long; she shrinks affrighted from the rude touch of the Sun and closes her petals during the day. The outer floral leaves of the Lily are green, and in the day-time the closed flowers are hardly noticeable among the broad green leaves which float on the water. In the evening, the scene is transformed as if by magic, and myriads of glistening white flowers cover the dark water (fig. 64). This phenomenon, recurring every day, has not only been observed by the poets, but an explanation has been offered for it: the Lily loves the moon and is frightened by the sun!

Had the poet taken out a lantern in the dark night, he would have noticed that the Lily opened even in the total absence of light from the moon! But the poet is not expected to carry a lantern and peer about in the dark; that inordinate curiosity is characteristic only of the man of science. The Lily, on the other hand, does not close with the appearance of the sun, for the flower often remains awake up to eleven in the forenoon.[1]

A French dictionary-maker consulted Cuvier, the zoologist, upon his definition of a crab as 'a little red fish which walks backwards.' 'Admirable!' said Cuvier; 'only the crab is not necessarily little, nor is it red till

[1] The Water-Lilies of Europe close at night and open in the day-time. The cause of this difference will be explained in the course of the chapter.

boiled; it is not a fish, and it cannot walk backwards; with these exceptions, your definition is perfect.' And so with the poet's description of the movement of the

FIG. 64. Nymphæa, closed and open.

Lily; it does not open to the moonlight nor yet close to the sun.

The sleep and waking of the Water-Lily is by no means an isolated occurrence. My attention was first drawn to another remarkable floral display by the folksong which begins:

> Our day's work is over,
> Like life's span but an hour!
> For now behold the gold-starred fields
> Of opening *Jhinga* flower.

Now I witness, every evening, a glorious transformation in my experimental garden at Sijbaria on the Ganges. The gardener has planted a large plot with *Jhinga* (*Luffa acutangula*). The flowers when closed in the day-time are very inconspicuous, the outer floral leaves being dull green; in my afternoon walk, I can hardly recognise the old familiar field; yet a little later, it is covered with masses of flowers in their golden glory. They remain open throughout the night, but close early in the morning; the fairy field of cloth of gold seems suddenly to have vanished.

We will now try to understand the phenomenon of the opening and closing of the Water-Lily. Numerous environmental stimuli act upon plants, and one of these is the action of light, of which the familiar example is furnished by the turning of the Sunflower towards the light. It has been found that of all the rays present in white light which are effective in producing the movement of plant-organs, the blue and violet rays are the most potent; the yellow and the red are practically ineffective. Now the moonlight, besides being feeble, contains but little of the effective rays; hence the light of the moon cannot cause the movement of the floral leaves.

The only effective light is that of the sun, but the opening and closing of the Water-Lily has little connection with the rising or setting of the sun. The opening cannot be due to the setting sun, for the flower is open in the forenoon; neither can it be due to the rising sun, since the flowers are already open when it rises. The daily movements of the flower cannot, therefore, be due to the alternating action of light and darkness.

Complexity of the Problem

The plant is affected not merely by a single form of stimulation. The phenomena of plant-movement, as previously stated, have remained obscure on account of the numerous factors which contribute to induce them. This will be better understood if we consider only two out of numerous factors operating on the plant—the stimulus of gravity and that of light. Certain organs are highly sensitive to geotropic stimulation, while others are feebly sensitive to it. The stronger reaction will be represented by G and the feeble by *g*. In regard to light, there are two distinct classes of effect: positive heliotropism when the organ turns towards light, and negative heliotropism when the organ turns away from it. Their effects when strong will be represented by $+\mathrm{L}$ and $-\mathrm{L}$; when feeble, by $+l$ and $-l$.

What will be the resulting effect when a horizontal stem is exposed to combined geotropic and heliotropic stimulation? In geotropic response the stem will bend upwards; should the organ be positively heliotropic, the curvature under vertical light will also be upwards. Geotropism and heliotropism will thus conspire, the joint effect being $\mathrm{G}+\mathrm{L}$; but should the organ be negatively heliotropic, the resultant will be $\mathrm{G}-\mathrm{L}$. If further account be taken of the relative sensitiveness of the organ to the stimuli of gravity and light, we shall have the following possible combinations:

$$\mathrm{G}+\mathrm{L};\ \mathrm{G}-\mathrm{L};\ \mathrm{G}+l;\ \mathrm{G}-l.$$
$$g+\mathrm{L};\ g-\mathrm{L};\ g+l;\ g-l.$$

Eight different effects can thus be produced by the combination of only two factors; there are, however, other factors present, such as the effects of rise and fall of temperature. Additional complications are introduced by the unequal sensitiveness of the two sides of the organ; in some, it is the upper side, in others it is the lower side that is more excitable and therefore reacts more effectively. There are thus at least ten factors in operation, and the different combinations possible would exceed a thousand.

It is no wonder that the movements of plants appear so extraordinarily complex. Efforts to discover the true explanation have long been baffled by the fact that it has hitherto been impossible to isolate and study the effect of each of the factors so as to analyse their complex result.

The Process of Elimination

I stated that the different possible variations of results produced by various combinations would exceed a thousand. Life would not be long enough to go through all of them one by one, on the chance of discovering the solution of a particular problem. It is possible, however, by simple preliminary tests, to eliminate the ineffective factors, and thus to bring the inquiry within narrow limits. Let us now consider the various factors which are likely to be present; these are the effects of stimulation by gravity and by light, and of variation of temperature.

Does the stimulus of gravity exert any marked effect on the movement of opening and closing of the flowers? The petals close up in the middle of the day, each of the petals standing erect. If the flower were susceptible to the stimulus of gravity, then, on turning the flower upside down, the closed petals in their inverted position would start to curl upwards and outwards, thus opening the flower. But no such effect takes place. We may next inquire whether variation of the intensity of light could induce the movement of the petals. Light appears in the morning and disappears in the evening. If the movement of the petals were entirely dependent on light, two opposite effects would be produced in the morning and evening respectively. But the flower is open at both these periods. Moreover, the course of the opening and closing movements does not strictly coincide with the daily change of light and darkness. The movement of the Water-Lily petals is, therefore, not essentially dependent on variations of light.

The Diurnal Record of the Water-Lily

It was shown that the movement of its petals was not affected by gravity nor by light. On what then does the movement depend? To ascertain this we have to take a continuous record of the movement of its petals. They are closed in the day-time; we want to know the exact moment when they begin their 'waking' movement, the time when this movement is most rapid and when the flower becomes fully expanded.

After this expansion, there must be a time for the reverse movement or closure of the petals. When does this begin, at what rate does it proceed, and when does the flower completely close its petals in 'sleep'?

The only other fluctuating element which may possibly affect the movement of the petals is the daily variation of temperature. In summer the minimum temperature is attained about 6 A.M. After this, the rise is very rapid, the maximum being attained at about 2.30 P.M. The temperature then begins to fall until the minimum is reached next morning. The time taken for a rise from the minimum to the maximum is 8½ hours; the period of fall through the same range is, on the other hand, 15½ hours. The rate of rise is therefore far more rapid than the rate of fall. In winter the minimum temperature is attained half an hour later and the maximum half an hour earlier. In order to determine the influence of temperature on the movements of the flower, it is necessary to obtain a daily record, both of the petals and of the change of temperature throughout the twenty-four hours.

By means of the Automatic Recorder the two records are taken simultaneously on the same plate. In the complete record (fig. 65) the upper curve shows the daily variation of temperature, and the lower curve the movement of the petal. The corresponding thick dots in the upper and the lower curves indicate the precise hour at which the two essential changes take place.

It will now be seen how astonishingly parallel is the curve of the movement of the flower to the curve which represents the variation of temperature. There can, therefore, be no doubt that the cause of the opening and closing of the flower is the diurnal change of temperature. The flower is in a position of sleep during the day; a rapid fall of temperature occurs from 6 P.M. and the petals begin to open, at first slowly, then very rapidly. The flower is completely open and fully expanded by 10 P.M. Though the temperature continues to fall, there is no further possibility of expansion beyond the maximum. At about 6 A.M. the temperature begins to rise, so the reverse movement of closure sets in. The flower continues to close very rapidly till the closure of 'sleep' becomes complete by about 10 A.M.

It is thus shown that the closing of the flower is brought about by a rise of temperature, the opening by a fall. The explanation of the movement is that a young flower is still in a state of growth, and rise of temperature accelerates while a fall of temperature retards it. Now in the petal of the Lily, the two sides differ in sensibility, just as we found the upper and lower sides of the pulvinus of Mimosa to be unequally sensitive. In the Indian Water-Lily it is the outer side which is the more sensitive. Hence, during rise of temperature, the outside grows faster than the

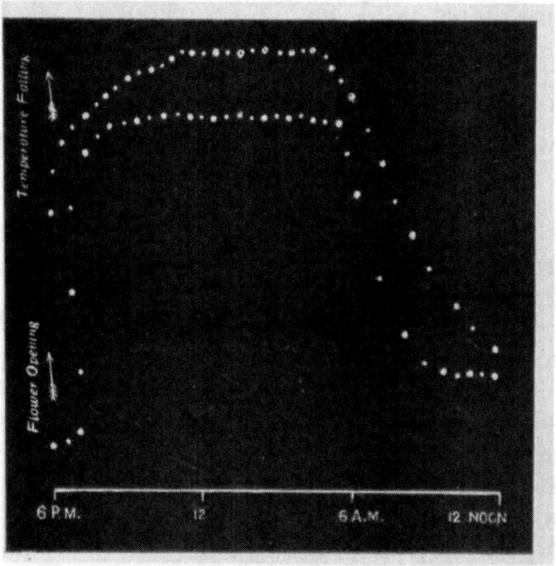

FIG. 65. Diurnal record of the petal of Water-Lily.

inside, thus producing a closing movement. During the fall of temperature, the reverse movement takes place, since the retardation of growth is greater in the more sensitive outer side.

In the European Lilies it is the inner side that is relatively the more sensitive. These flowers are compelled to reverse the order of the reactions of the Indian Water-Lily, by opening in the day-time and closing at night. They cannot, therefore, be accused of moon-worship. They have the healthy habit of normal humanity to sleep at night and keep awake in the day-time. Others turn night into day, and make up for their long night-watch by sleeping it off in the day-time!

MOVEMENT OF CASSIA UNDER VARIATION OF LIGHT

As an example of an organ specially sensitive to light we may take the leaflet of the Indian plant *Cassia alata*.

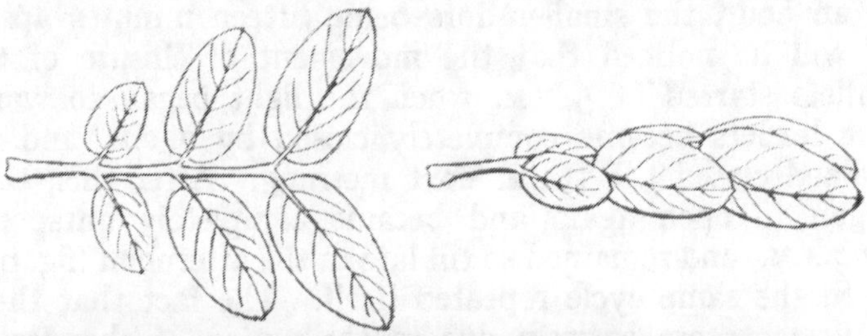

FIG. 66. Leaf of Cassia in open and closed positions.

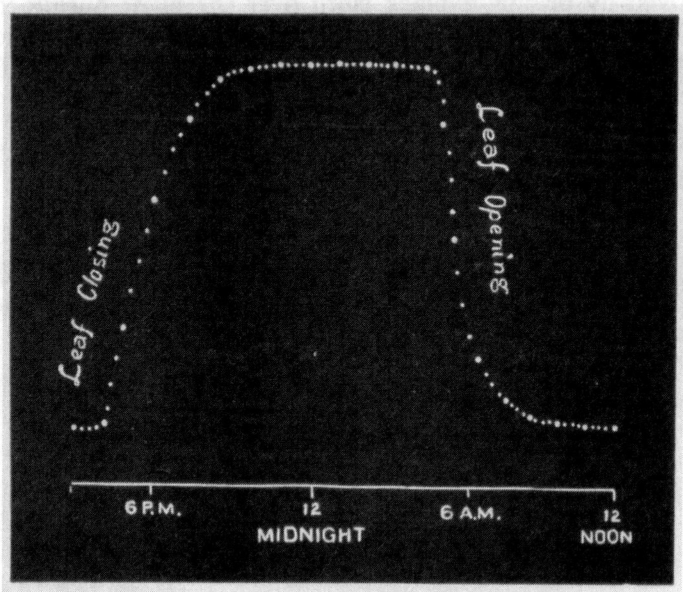

FIG. 67. Diurnal record of a leaflet of Cassia.

These leaflets remain tightly closed during the night, but from early morning onwards they begin to open and remain widely spread out throughout the day (fig. 66). So sensitive are the leaflets that cloud-shadows cause them to begin to close.

In order to show that the daily opening and closing movements of this plant are entirely due to alternation of light and darkness, a continuous record was taken from 4 P.M. till noon the next day. The first thick dot was marked at 4 P.M.: the successive thick dots are at intervals of an hour, the smaller dots being fifteen minutes apart. It will be noticed that the movement of closure of the leaflets started at 5 P.M. when the light began to wane. The leaflets became completely closed by 9 P.M. and remained closed till 5 A.M. next morning. After this, they began to open again and became completely outspread by 9 A.M., and remained so till late in the afternoon (fig. 67). Then the same cycle repeated itself. The fact that these movements are entirely due to the action of changes in light and not to variation of temperature can easily be proved by placing a black cloth over the plant during the middle of the day. The temperature remains the same; yet because of the artificial darkening, the leaflets rapidly close.

CHAPTER XVII

THE RISE OF THE SAP

THE plant absorbs from the soil water holding food-material in solution. The distribution of the absorbed liquid, that is the rise of the sap, enables the plant to maintain the cells in that state of turgor without which the growth of the plant and its various life-movements would be arrested. But how is the sap raised to the top of a tree?

This question has perplexed scientific investigators for more than two hundred years; it has been argued, on the one hand, that it is wholly due to the action of physical forces, and on the other, that it is brought about by some activity of the living tissues. For reasons to be presently explained, the actual consensus of opinion is in favour of the physical theory, but the evidence to be here adduced proves that view to be quite untenable.

Let us take the instance of a tall tree with numerous leaves. What are the physical forces which may be concerned in raising the sap from the roots to the leaves? Transpiration, that is the exhalation of water-vapour, is constantly taking place from these leaves and a partial vacuum is produced in the wood-vessels which run through the plant from top to bottom. The atmospheric pressure will consequently force the water up the plant; but the greatest height to which water can be raised in this manner is only 34 feet, the height of the water-barometer. The Palms, however, grow to a height often exceeding 100 feet; but even the Palm is a pigmy compared to some of the giants like *Eucalyptus amygdalina* with a height of 450 feet. The theory of atmospheric pressure is, therefore, inadequate, as is also that of capillarity.

The theory of osmotic action has also been invoked in explanation of the phenomenon. When a semi-permeable bag containing a strong sugar-solution is dipped in water, the water passes into the bag containing the strong solution. The vegetable cells are like so many bags containing strong solutions, and they will, therefore, take up water accessible to them. The cells in the root thus take up water from the soil, the next cell higher from the one below it, and so on in progressive series upwards.

But this osmotic process is exceedingly slow. Let us see what this means when the leaves at the top of a Eucalyptus are on the point of death from severe drought. We can well imagine the impatience with which the parched leaves would be waiting for the sap to reach them after rain had irrigated the roots. If the water-movement depended solely on osmotic action, then their prospect would be none too cheerful, for the shortest time in which water would reach the top by osmosis would be considerably over a year! Among the upholders of the physical theory is Strasburger, an eminent plant-physiologist; but even he has been constrained to say that 'Osmotic forces act too slowly to be of any value, and, moreover, there is no fixed distribution of osmotic substances that would account for such a current.' [1]

There must, therefore, be some other agency at work to effect a more rapid propulsion of the sap.

Pull and Push by the Terminal Organs

Another theory is that the rise is due to the action of a pull from above and a push from below exerted by the terminal organs, the leaf and the root. This theory is only partly physical, for both the pull and the push depend upon the vital activity of leaf and root respectively. The removal of water from leaves by transpiration is supposed to exert a pull along cohering columns of water in the vessels present in the wood. But the water-columns in the vessels are not continuous but are interrupted by air-bubbles; it is obviously impossible to employ ropes of water for haulage, particularly if they are severed here and there! So much for the pull from above. The push from below is supposed to be exerted by root-pressure. But in the Palm there is no detectable root-pressure, yet the sap rises

[1] Strasburger, *Text-Book of Botany* (English Translation, p. 187).

to a height of more than a hundred feet. Again, during active transpiration, when the need of the tree is greatest, the root-pressure, instead of being positive, is actually negative. The fact that neither the leaf nor the root is absolutely necessary for the propulsion of sap will be demonstrated later, when I show that the sap-movement persists even after complete removal of the root and the leaves.

Propulsion of Sap essentially due to Vital Activity

The various physical theories having proved unsatisfactory, there remains the question whether the activity of living cells might not be instrumental in effecting the propulsion of sap. This physiological theory, however, received a severe blow from some inconclusive experiments of Strasburger, which led him to the conclusion that the ascent of sap is not affected by the application of poison which must necessarily kill all living tissues. My results, to be presently described, lead to a conclusion diametrically opposite to Strasburger's.

The following experiments on the effect of stimulants and poisons are quite decisive in proving that the propulsion of sap is essentially due to the activity of living cells. A cut stem of *Chrysanthemum coronarium* is subjected to drought; the plant doubles over, the leaves shrink and crumple up; in fact, the plant seems dead. But irrigation with water containing traces of a stimulant brings about a marvellous transformation; there is now an active rise of sap by which the original turgor is restored. The bent stem straightens up and the leaves spread out in their original vigour, as seen in the photograph reproduced (fig. 68); complete recovery was, in this case, found to take place in as short a time as fifteen minutes. In sharp contrast with this was the parallel experiment, in which the drooping stem was irrigated with poisonous formaldehyde solution. The specimen never recovered, but underwent a complete collapse into a huddled mass of dying tissue (fig. 69).

I introduced certain modifications in the next experiment; the specimen employed was also different. An erect stem of *Centaurea* was placed in a vessel P containing a poisonous solution of potassium cyanide, while a drooping stem of the same plant was placed in the vessel S, which contained a stimulant. The opposite effects of poisonous

FIG. 68. Full erection of a cut shoot with drooping leaves on application of water at the cut end. (*Chrysanthemum.*)
Photographs showing the plant before (right) and after irrigation (left).

and stimulating solutions are strikingly shown in the pair of photographs to the right. The erect specimen under poison shows a complete collapse, while the drooping stem under stimulant exhibits a vigorous recovery (fig. 70).

The evidence adduced suffices to prove that the propulsion of sap is essentially due to the activity of living tissues. But a vague assumption of living activity is not a complete explanation of the phenomenon. It is necessary further to determine the character of the underlying activity,

FIG. 69. Photographs of drooping cut shoot of *Chrysanthemum* placed in solution of Formaldehyde, which caused increased drooping, instead of full erection due to ascent of water as in fig. 68.

FIG. 70. Erect stem placed in poisonous and drooping stem in stimulating solution (left).

The poisoned stem droops and dies, while the drooping stem becomes erected (right).

how that activity is initiated, and by what means a definitely directioned transport of sap is maintained.

In order to obtain a complete solution of this problem, I had to devise diverse methods of experimentation and various instruments. The results obtained led to the conclusion that the sap in the plant is propelled by a mechanism which is essentially similar to that which maintains the circulation of blood in the animal.

Propulsion of Blood in the Animal

The flow of blood in a higher animal is maintained by the automatic and ceaseless pulsations of a contractile organ which we call its heart. We must, however, adopt a generalised idea of the heart and its mechanism, for it would not be reasonable to seek in the plant for an organ so highly complex and centralised as, say, the human heart. The correspondence with the plant is rather to be sought in the lower animals, in which, as in Amphioxus, the heart is an elongated tubular organ, the contained nutrient fluid being propelled by waves of peristaltic contraction which sweep forward. In the embryo of even the higher animals, the heart is an elongated tube. The essential characteristic of the cardiac tissue of the animal is its rhythmic pulsation, the rapidity of which is appropriately modified under definite conditions.

Thus under certain stimulating drugs the heart beats faster, and pumps the blood at a quicker rate, ensuring a more rapid flow; a depressant induces a precisely opposite reaction.

A certain amount of internal tension is necessary to initiate the pulsation of the heart; thus the quiescent heart of a snail is made to pulsate by subjecting it to an increased intracardiac pressure.

The activity of the heart is increased within limits under a rise of temperature, and decreased by application of cold.

A small dose of a narcotic like ether accelerates the cardiac activity. A strong narcotic like chloroform causes an immediate stimulation followed by depression and arrest under its continued action.

I will now show that any condition which enhances the activity of the heart and thus increases the rate of flow of blood, also increases the rate of flow of sap; and conversely, that depressing agents lower or arrest the flow in both. The difficulty to be overcome was the fact that no satisfactory method had hitherto been available for the detection and measurement of the normal rate of ascent of sap and its induced variations.

The Leaf as an Indicator of the Flow of Sap

I succeeded in removing the difficulty by utilising the leaf as an indicator of the movement of sap. In a potted plant, drought is found to produce, sooner or later, a drooping of the leaves, whereas after irrigation they become erect. Moreover, any agent which increases the activity of pumping the sap, also causes a rapid erection of the leaf; diminution of the pumping activity by a depressant produces, on the other hand, a fall of the leaf. These movements are, however, too slight to be easily observed; they become conspicuous when highly magnified, as by the following device.

The Electric Phytograph

This is a very sensitive apparatus for the detection of changing rates in the ascent of sap. The leaf is attached by a thin silk thread to the recording lever, which magnifies the movement of the leaf from ten to a hundred times. The record is taken on a smoked-glass plate. The lever, by means of an electro-magnetic device, is made to tap successive dots on the moving recording plate at intervals of two to ten seconds (fig. 71). The response obtained with the leaf of a cut shoot is practically the same as that of an intact plant with roots; a cut shoot obviously offers greater facilities for manipulation.

Two small test-tubes filled with different solutions are mounted on a revolving rod, which can be raised or lowered by means of a handle. The cut end of the shoot may thus be acted upon in rapid succession by a chemical depressant or a stimulant, or by cold or warm water.

To demonstrate the similarity of action of the propulsive mechanism in plant and animal, I will give a detailed description of the opposite effects produced by depressing and stimulating agents. The leaf was in a balanced hori-

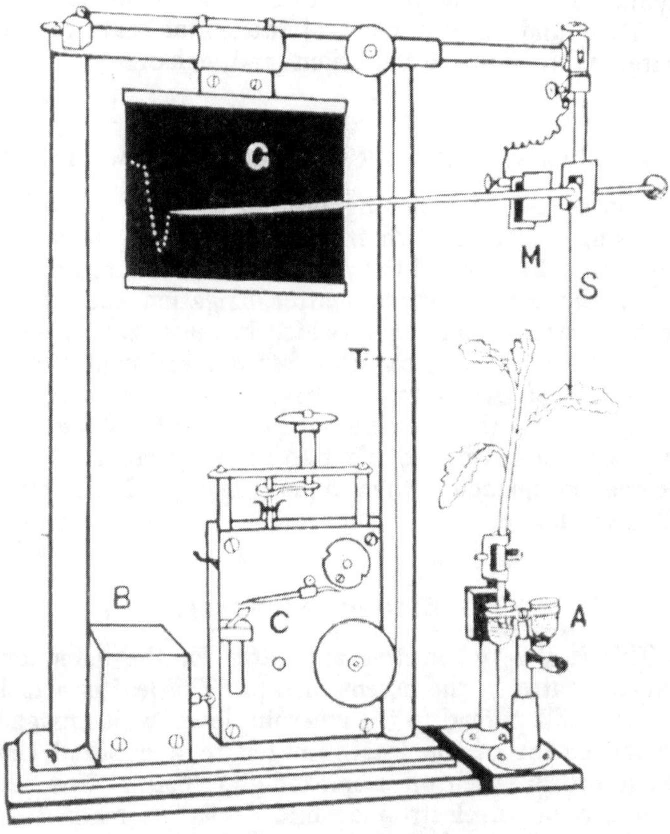

FIG. 71. The Electric Phytograph.

zontal position. Application of potassium bromide solution produced so great a depression that the leaf exhibited a rapid fall. The effect of the application of a small dose of camphor, which is a stimulant, was the arrest of the fall and the subsequent rise of the leaf (fig. 72). Hence it may be inferred that the supply of sap to the leaf was diminished in the one case and increased in the other, and that the reagents must have affected living cells by which the supply is maintained.

The effect of drought was exhibited by the fall of the leaf. Here again the propulsion of sap must have been interfered with, probably by a diminution of the internal hydrostatic pressure in the propulsive cells.

The effect of alternate application of cold and warm water in successive depression and enhancement of the

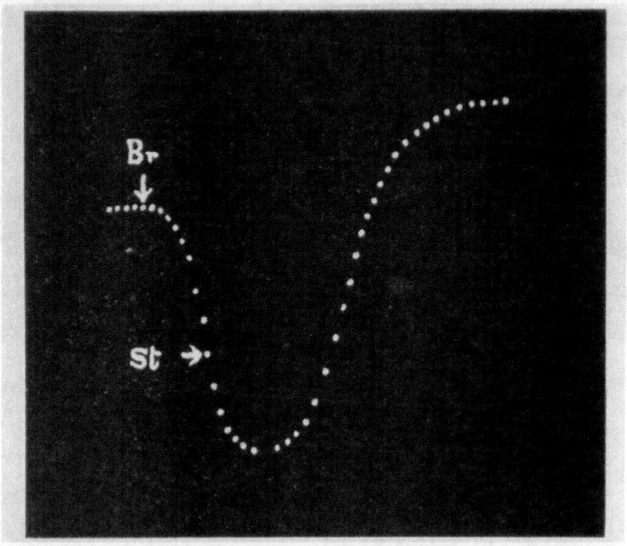

FIG. 72. Alternate arrest and enhancement of ascent of sap by chemical depressant and stimulant.

rate of ascent of sap was clearly demonstrated in the record (fig. 73).

Dilute ether was found to produce an enhancement in the rate of ascent and a rapid erection of the leaf. Stimulation followed by depression under the action of chloroform was exhibited by the preliminary erection of the leaf followed by a rapid fall.

Some of the important facts established in this chapter may now be recapitulated. The arrest and permanent abolition of the ascent of sap in a plant treated with poison prove it to be due to the activity of living tissue. The alternate arrest and renewal of the ascent under cold and warmth, and the characteristic enhancement or depression of the rate under cardiac depressants and stimulants, further

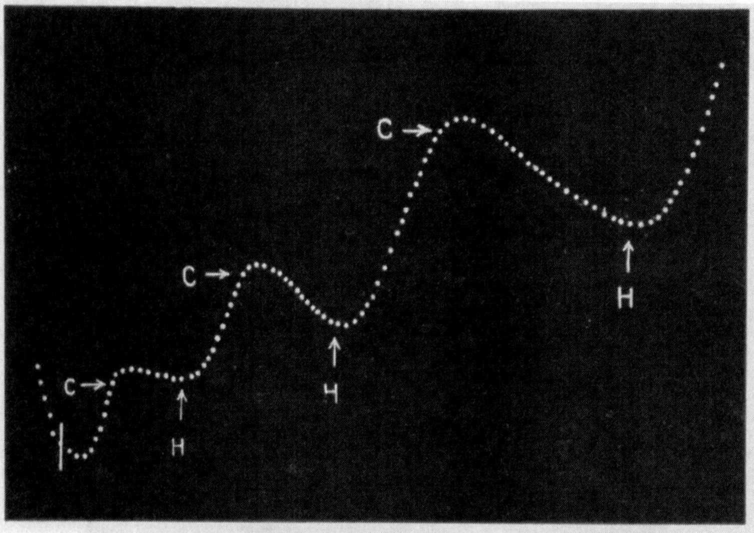

FIG. 73. Effect of alternate application of cold (C) in depression (down-curve) and of warmth (H) in enhancement (up-curve) of rate of ascent.

show that the mechanism for the propulsion of fluids is fundamentally similar in plant and animal. Physical mechanism could not have in any way manifested these characteristic reactions.

CHAPTER XVIII

THE PROPULSIVE TISSUE

THE experiments described in the last chapter prove that there is an active tissue somewhere in the interior of the plant, the pulsation of which effects the propulsion of the sap, just as the pulsation of the heart maintains the circulation

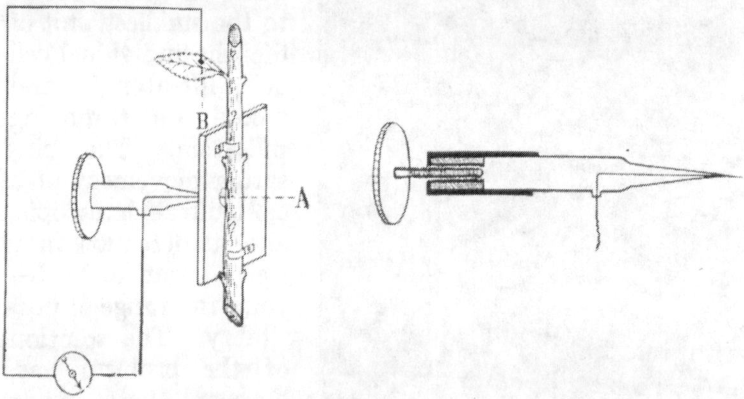

FIG. 74. The Electric Probe for the localisation of the Pulsating Layer.

The point of the Probe enters the stem at A, the second electric contact being made with a distant leaf. The figure to the right is an enlarged view with the micrometric screw for the gradual introduction of the Probe into the tissues of the plant.

of the blood in the animal. There must, therefore, be something like a primitive 'heart' in the plant, but not so centralised and highly differentiated as in higher animals. In the lower types of animal, as also in the embryo of the higher, the heart is an elongated organ, the contained nutrient fluid being propelled forward by peristaltic contraction. The propulsion of sap in plants, I find, is due to a similar peristaltic action. As the pulsating mechanism in plants is not so highly differentiated as in higher animals, the propulsive organ is likened to an elongated 'heart,' using that term in a broad sense. The propulsive system of plants may be said to correspond to the heart and arteries of animals.

SEARCHING FOR THE 'HEART' OF THE PLANT

Where then is the 'heart'? Is it possible to localise exactly the active cells in the interior of the tree and catch them, as it were, in the very act of pumping up the sap? In attempting this, we have to gain access to the smallest unit of life, the individual cell or 'life-atom,' and record its throbbing pulsation. The pulsating movement of a cell is ultra-microscopic, and its detection may well appear to be beyond the range of possibility. The solution of the problem has, however, been made possible by my Electric Probe, previously employed for localising the layer which perceives the stimulus of gravity.

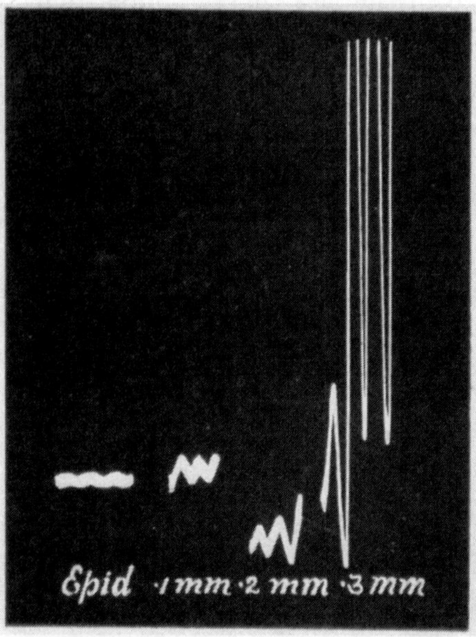

FIG. 75. Record showing the amplitude of electric pulsation at different layers.

Note the abrupt enhancement at a distance of 0·3 mm. from the surface, the particular layer being in the inner cortex; a portion of the record has gone out of the plate.

The location of the 'heart' was effected by the employment of the Electric Probe in circuit with a sensitive galvanometer. When an electric contact is made with a resting muscle, the galvanometer remains quiescent. But if the contact be made with a beating heart, electric pulsations are generated corresponding to the mechanical pulsations. In locating the 'heart' I introduced the Probe, step by step, into the organ; as soon as it came in contact with the pulsating layer, electric signals were sent out, which were automatically recorded in the galvanograph.

The experiment was carried out as follows: one electric contact was made with a resting tissue, such as the epidermis of the leaf; the other contact was made by the Probe (fig. 74), which was introduced transversely into the stem

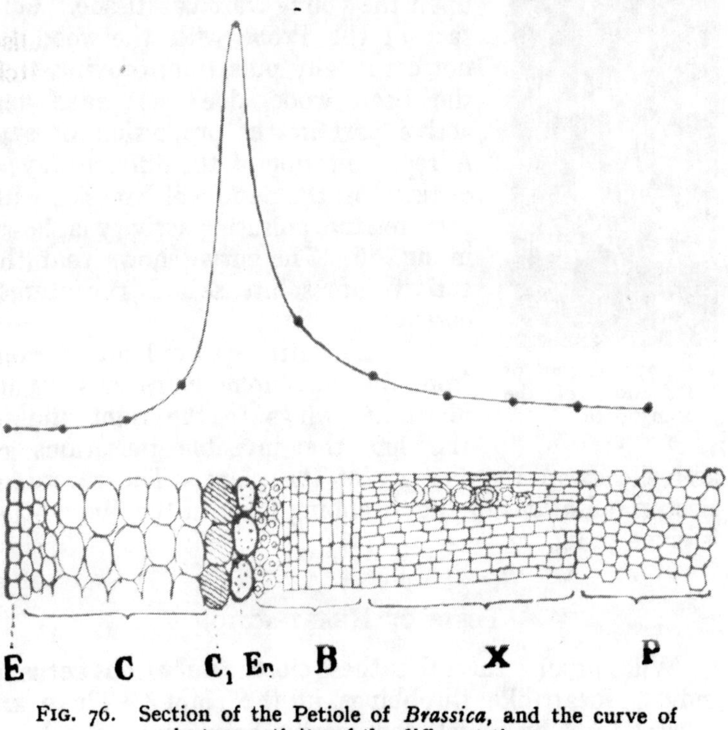

FIG. 76. Section of the Petiole of *Brassica*, and the curve of pulsatory activity of the different tissues.

E, epidermis; C, cortex; C_1, the active internal cortical layer; E_n, endodermis; B, phloem; X, xylem; P, pith. Note the sudden enhancement of activity at the layer C_1.

by successive steps of 0·1 mm. No pulsation could be detected at the epidermis; as the Probe reached a depth of 0·1 mm. a feeble pulsation was detected; a similar result was obtained at the depth of 0·2 mm. But when the Probe reached 0·3 mm. the pulsations exhibited a sudden increase; this was so great that a part of the record went off the plate (fig. 75); evidently the Probe had come into contact with pulsating cells. As it was thrust still deeper into the stem, the pulsating activity rapidly disappeared. When a transverse section of the stem was made along the line of the passage of the Probe, it was found that the maximum activity had been detected when the Probe touched the internal layer of the cortex abutting upon the young vascular tissue. Contact of the Probe with the wood did not cause any pulsation, proving that the dead wood does not take any active part in the propulsion of sap. A representation of the different layers of tissue in the petiole of Brassica with their relative pulsating activity is shown in fig. 76. The curve shows that the activity is most intense at C_1, the internal cortex.

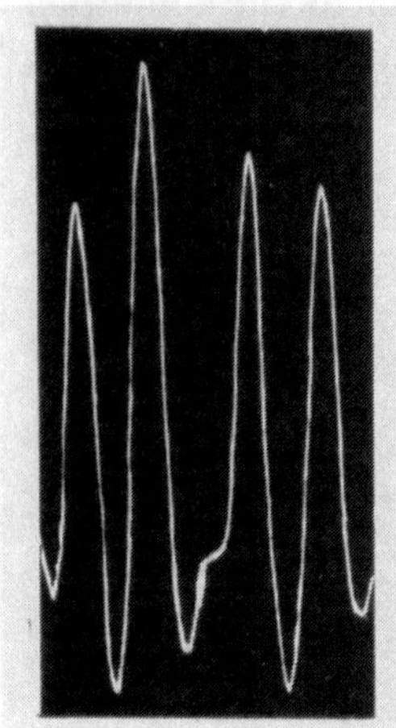

FIG. 77. Record of Pulsations of the Mango-tree.

The indicating spot of light reflected from the galvanometer reveals by its alternate swings to the right and to the left the invisible pulsations of active cells in the interior of the plant. The record of electric pulsations of the throbbing layer in the Mango-tree is given in fig. 77.

TESTS OF HEART-ACTION

What proof is there that these electric pulsations actually indicate heart-like throbbings in the plant? There are several tests by which the specific activity of the heart may be discriminated, of which the following will suffice: (1) The heart-beat comes to a stop when the internal blood pressure is low, the pulsations being renewed after an increase of internal pressure. (2) The beat also comes to a stop when the heart is in a depressed or sub-tonic condition; stimulation is then found to revive the beating of the heart. (3) Anæsthetics like chloroform stimulate the heart at the beginning, but under prolonged anæsthetisation it comes to a stop, followed by the death of the animal. I shall now show that the pulsatory reactions of the plant are exactly parallel to those of the animal.

Effects of Diminished and Increased Internal Pressure

The sap-pressure undergoes a great diminution under drought, in consequence of which the pulsation of the propulsive layer comes to a stop. An increase of sap-pressure after irrigation revives the arrested pulsation. I reproduce the effects of alternate withholding of water and of irrigation

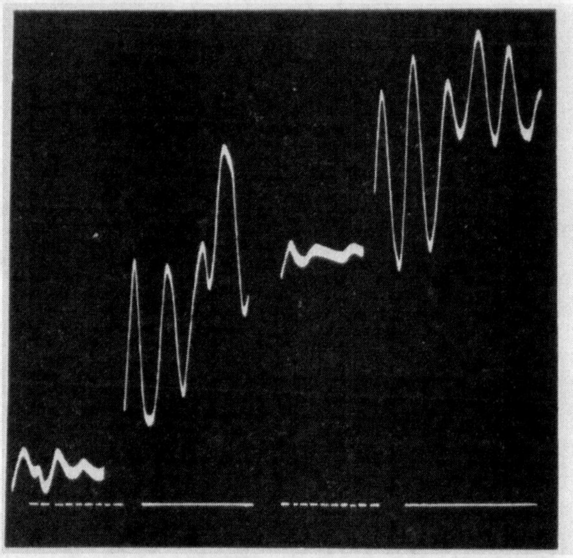

FIG. 78. Records of electric pulsations under alternate drought and irrigation.

Dotted line below represents condition of drought, continuous line of fresh irrigation.

on the electric pulsation of the propulsive layer. The pulsation is seen to be repeatedly depressed under drought and revived after supply of water (fig. 78).

Effect of Sub-tonic Condition

When a plant is kept in the dark for twenty-four hours it becomes so sub-tonic as to lose its power of maintaining the ascent of sap. Stimulation by light or by electric shock is now found to revive the activity of ascent. What can be the fundamental cause of this change? The explanation is to be found in the records of the electric pulsations. These show that, while the pulsations are practically at

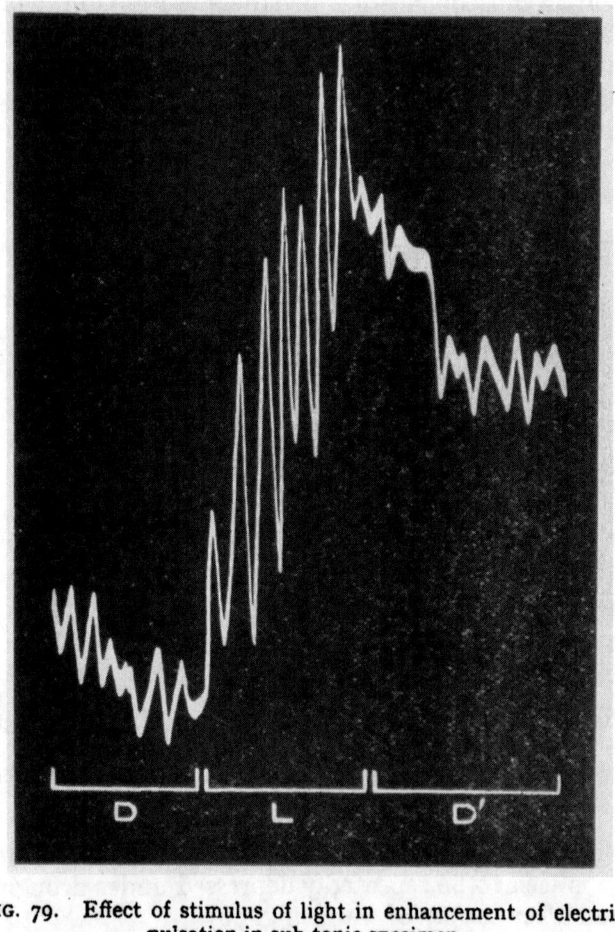

FIG. 79. Effect of stimulus of light in enhancement of electric pulsation in sub-tonic specimen.

D, feeble pulsation in dark; L, enhanced pulsation under light; and D′, depression under renewed darkness.

a standstill under prolonged darkness, the throbbing activity is revived on stimulation by electric shock or by light, and is arrested once more on the cessation of the stimulus (fig. 79).

Effect of Anæsthetics

It has been explained how, in the first stage of its application, chloroform enhances the activity of the heart, continued action producing depression and arrest. I have also found that the rate of sap-ascent under chloroform shows a preliminary acceleration followed by depression and arrest.

The record of the electric pulsations of the propulsive layer under chloroform shows parallel reactions. The preliminary enhancement is followed by depression and arrest of pulsation (fig. 80).

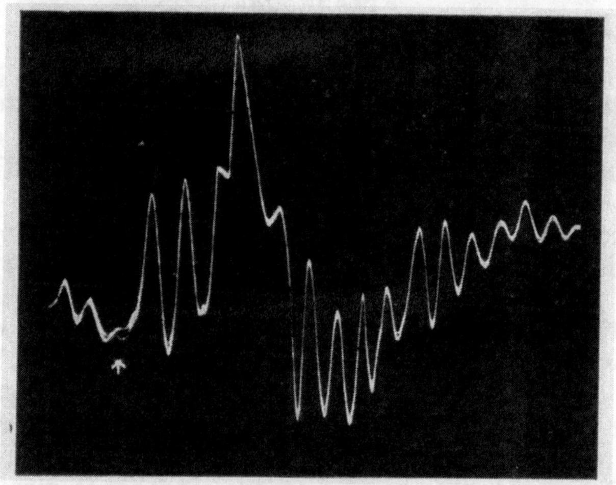

FIG. 80. Effect of Chloroform, applied at arrow, on pulsation.

Note preliminary enhancement of pulsation, with longer upstroke; pulsation arrested on continued application.

The results described offer conclusive proof that the ascent is brought about by the pulsating activity of the cells of the propulsive layer, which is the inner cortex surrounding the vascular cylinder. It should be borne in mind in this connection that, under exceptional circumstances, all living cells are capable of being thrown into rhythmic activity; certain layers of cells are, however, naturally more active than others, and it is by the pulsatory activity of those of the inner cortex that the normal ascent of sap is maintained.

The propulsive tissue in a Dicotyledonous-tree is a cylindrical tube which stretches throughout its entire length. This cylinder closely surrounds the young vascular tissue. The function of this long cylinder, as already explained, is not unlike that of the elongated heart of lower animals, in which circulation of the blood is brought about by a series of peristaltic waves. The propulsion of sap in plants is found to be essentially a similar process of peristalsis, in which a wave of contraction squeezes the sap forward. A succession of such peristaltic waves maintains the continuous ascent of the sap.

Normal and Reversed Peristaltic Waves

It may next be asked why should the peristaltic wave always move the sap upwards. Would it be possible to reverse the direction of the peristaltic wave, so as to cause the sap to flow downwards? What are the conditions which determine the direction of flow?

The general law which I have established concerning the direction of the sap-movement is *that it flows from the more active to the less active region.* It is evident that if the pulsating activity were equal at the two ends of the organ they would balance each other, and there could then be no resulting directive movement. A difference of activity at the two ends of the organ may be induced in two ways: first by differential turgor, and second by differential stimulation. I now proceed to explain what is meant by differential turgor and differential stimulation.

After irrigation the lower end of a tree is rendered tense and turgid by absorption of water, whereas there is an incipient drought at the upper end in consequence of rapid removal of water by the transpiring leaves. Now the rhythmic activity of the propulsive cells has been shown to be enhanced under increased turgor and diminished under drought. The propulsion of sap is, therefore, from the more turgid and active to the less turgid inactive region. The sap-movement thus follows the 'turgor-gradient,' tending to equalise the degree of turgor in different parts of the plant.

I shall next show that it is possible to change the normal upward direction of the flow of sap into a downward one, by merely reversing the turgor-gradient. Thus when water is withheld from a potted plant, the stem bends over under drought and the drooping leaves hang down. The pulsating activity throughout the plant is now in a state of arrest. If a glass of water be next raised so that the upper part of the drooping stem is immersed in it, then the upper part of the stem will absorb water and thus become more turgid than its lower end. The pulsating activity becomes revived at the upper end; the turgor-gradient is reversed, and the sap now flows downwards against the direction of the normal ascent. This reversed flow is demonstrated by the sequence of revival of the drooping leaves which takes place from the tip of the stem downwards. I have measured the normal rate of ascent and the reversed rate; the results show that the rate is much slower in the reverse or unaccustomed direction. In the circulation of the blood the forward propulsion is helped by the presence of valves which facilitate the flow in one direction rather than its opposite. The partitions of the cellular pumps in the plant act somewhat similarly in predisposing the flow in the normal upward direction.

Under normal conditions the root-cells are continuously stimulated by friction against the soil, and this probably starts the peristaltic waves. The increased turgor of the lower end under irrigation also directs the propulsion of sap in an upward course from the more to the less active region.

Function of the Wood

In herbaceous plants the distance of the leaves from the soil-water is not too great, but in tall trees it is necessary to have a nearer source of supply—a 'soil extension,' as it were, in the shape of conduit-pipes filled with water. These conduit-pipes are the young wood-vessels (alburnum) which serve for the mechanical transference of water during the emergency of active transpiration by the leaves. When transpiration is feeble, the normal ascent along the cortex supplies every portion of the tree with water; the leaves become turgid and the alburnum filled with sap. During active transpiration, however, the physiological propulsion is not sufficient to meet the demand, and water is withdrawn from the wood-reservoir. Two factors are thus brought into operation—physiological propulsion by and along the active cortical cells, and physical transference along the wood or xylem.

We may now visualise the important processes connected with the ascent of sap. The absorbing root-cells are continuously stimulated by mechanical friction against the soil, giving rise to peristaltic waves of pulsation along the active propulsive layer of the inner cortex. The direction of the sap-flow is determined by differential turgor, from the more turgid lower part of the plant to the upper part which is in a state of incipient drought due to active transpiration by the leaves. The rhythmic contraction of the active cells propels the sap not only upwards but also laterally into the young xylem, which functions as a reservoir for emergencies, water being withdrawn from it when the transpiration is most active.

CHAPTER XIX

THE WEEPING MANGO-TREE

SINCE the conclusion of my investigations on the causes underlying the supposed devotional exercises of the famous Praying Palm of Faridpore, I have been embarrassed by many requests to interpret happenings which, at first sight, appeared supernatural. The unexpected occurrence, in the present case, was the periodic 'weeping' of a Mango-tree in the suburbs of Calcutta.

This tree is full grown and about 40 feet in height. The circumference of the trunk is 38 inches and the outspread branches with their numerous leaves cover an area of about 100 square yards (fig. 81). The so-called 'weeping' commences every day without any ostensible provocation punctually at 1 P.M. from a point high up in the tree. It is very copious at the beginning, the rate of fall of successive drops being one in every two seconds. The paroxysm gradually subsides, and the interval between successive drops slows down to five seconds at 2 P.M., to eight seconds at 3 P.M., to fifteen seconds at 4 P.M., to 150 seconds at 5 P.M., after which there is a cessation of all 'weeping' for the day. The performance, moreover, is repeated every day at 1 P.M., and with the same sequence. The mysterious event came to be regarded as of evil omen, and thus aroused considerable alarm among the people in the neighbourhood. In their perplexity they asked me to explain what it meant, and if possible to cure the tree of its distressing symptoms.

It appeared, at first sight, that the pressure of sap inside the tree became somehow suddenly increased at 1 P.M., and that the sap then forced its way out through some vent high up in the tree.

Hourly Variation of Sap-pressure

When the internal sap-pressure in the trunk of a tree is very great, sap exudes from a hole drilled into it. The internal pressure may, on the other hand, be lower than

Fig. 81. Photograph of the 'Weeping' Mango-tree.

Exudation takes place through the small aperture marked with an arrow. The pressure-recorder is seen attached to the tree.

the external pressure of the atmosphere. Instead of there being any exudation, water is then actually sucked into the hole. The change of internal pressure, either increase or decrease, can be continuously recorded by attaching to the tree-trunk a self-recording pressure-gauge.

The pressure at any hour depends on the relative gain or loss of water by a tree. Water is absorbed from the soil and is forced up the trunk by the pumping mechanism as already described. The activity of this is increased, within limits, with the rise of temperature. The moment of the attainment of the highest temperature I designate as the *thermal noon*, that of the lowest temperature as the *thermal dawn*. The maximum temperature, under normal conditions, is attained in Bengal at about 2 P.M. and the minimum about 6 A.M.

Pressure Variation in Leafless Trees

There are many trees which shed their leaves about winter-time. There is then no loss of water by transpiration from the leaves, the variation of sap-pressure being determined by changes of the rate of suction under changing temperature. The maximum pressure is at thermal noon at 2 P.M., the minimum pressure at thermal dawn at 6 A.M. The sap escapes at noon with great rapidity from a hole drilled into the trunk.

Variation of Pressure in Leafy Trees

The conditions in trees with a large number of transpiring leaves are somewhat complicated. It is true that water is absorbed with greatest vigour at thermal noon, but loss from transpiration by the leaves is also greatest at that time; in fact, the loss is even greater than the gain. *Hence the diurnal variation of pressure in leafy trees is the reverse of that in trees without leaves.* The maximum pressure in plants with leaves is attained at thermal dawn, and the minimum at thermal noon.

In fig. 82 the upper curve shows the variation of temperature from 6 A.M. to 6 P.M. The lower curve shows the variation of pressure in a leafy Rain-tree (*Pithecolobium*). The minimum pressure is seen to occur at the thermal noon about 2 P.M. The pressure-curve appears to be more or less an inverted reflection of the curve of temperature. The highest temperature at noon corresponds to the minimum internal pressure. Hence at noon water is sucked in and not exuded at a drilled hole.

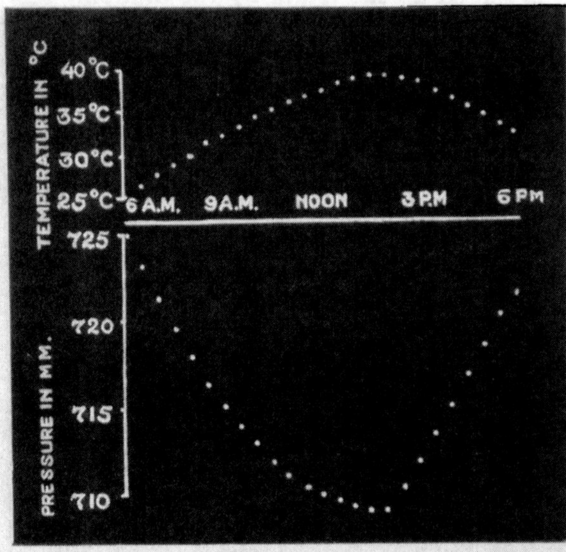

FIG. 82. Diurnal curve of variation of pressure in the Rain-tree (*Pithecolobium*).

The upper curve shows the variation of temperature, and the lower the variation of pressure. Note that the pressure was at its minimum at thermal noon, 2 P.M.

EXUDATION FROM THE MANGO-TREE

As there were numerous transpiring leaves on the Mango-tree, the internal pressure should have been at its minimum at noon. Nevertheless its 'weeping' or exudation of sap was then at its maximum. Closer examination of the tree disclosed the fact that there was a small aperture or vent in the bark at a point high up in the tree. This vent was usually closed by a plug of mucilage till forced out by strong internal pressure exerted by the sap at 1 P.M. It was an anomaly that the maximum exudation should take place at about noon, when the internal pressure in leafy trees is at its lowest.

This led me to undertake investigations on the pressure-variation in the Mango-tree; a hole was drilled for the attachment of a recording manometer at a point diametrically opposite to the vent. The pressure-variation was found, as expected, to be at its lowest at noon; there was, moreover, not the slightest exudation from the drilled hole. What then could be the difference in the same zone of the trunk, which would account for active exudation from the vent on one side, and the total absence of it from the drilled hole on the opposite side?

I continued the inquiry by cautiously removing the bark and the tissue round the vent. This led to the discovery of a large elongated cavity of irregular shape, formed by the decomposition of the young wood (alburnum) which had fallen to the bottom. The outer boundary of the cavity was a rind consisting of phloem and cortex in a healthy condition. The inner boundary was formed of duramen or hard wood, which is a non-conductor of sap (fig. 83). The total absence of exudation from the drilled hole on the left side of the trunk, and the copious overflow from the cavity on the right side, can only be due to the structural difference between the two sides—the presence of alburnum on the left and its absence on the right. The active cortex bordered the cavity, and the excretion of the fluid that fills it can only be due to a lateral pumping action of the cortex, the activity of which is gradually increased by the rise of temperature which attains its maximum at 2 P.M.

We have next to explain the reason of the commencement of exudation punctually at 1 P.M. Further inspection of the tree showed that for the greater part of the day the leaves cast a shadow on the trunk. There was an opening, however, among the branches, such that during the course of the sun from the east to west, sunlight fell directly upon the exuding portion of the trunk exactly at 1 P.M., causing a local rise of temperature. Consequently, the cortex underneath had its activity greatly increased, and the resulting enhanced exudation caused a rapid rise of the level of the sap collected in the cavity. The great increase of pressure expelled the closing plug, with a resulting sudden outflow of sap. Late in the day the sun became hidden by the leaves and the temperature underwent a rapid fall. The weeping of the tree consequently declined and was arrested in the evening.

On the left side of the trunk the young wood or

alburnum was uninterrupted, and could readily absorb whatever water might be laterally injected into it by the propulsive cells of the cortex. There was, therefore, no

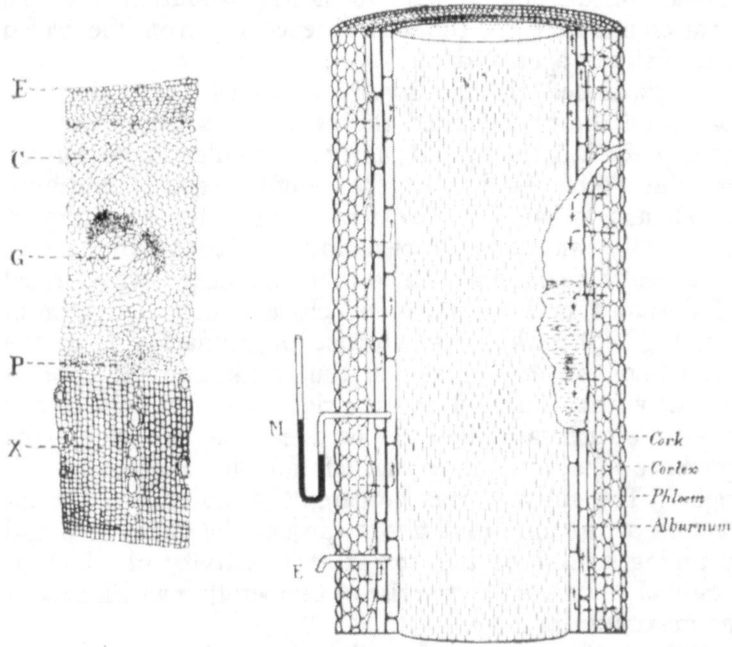

FIG. 83. Sections of the Mango Stem.

Figure to the left is a magnified transverse section of a young stem. E, epidermis; C, broad cortex; G, gland; P, phloem; X, xylem. Figure to the right is a diagrammatic representation of the trunk with the cavity from which exudation takes place. The sap pumped laterally by the cortex is accumulated in the cavity. On the left side the laterally injected sap is rapidly absorbed by the alburnum which is under negative pressure on account of transpiration from the leaves. The pressure was indicated by the manometer, M. No exudation took place through the drilled hole, E.

accumulation of sap and no exudation from the hole drilled on the left side.

The 'weeping' was thus due to the accumulation of laterally injected sap in the cavity and its periodic overflow from the vent. The tree wept no more after the destruction of the cavity and the coating of the exposed surface with coal tar.

The results given above furnish conclusive proof that the pulsatory activity of the cortex propels the sap not only upwards but also in a lateral direction, so as to inject it into the contiguous young wood-vessels which serve as a reservoir. It also proves that the young wood is a channel for the mechanical transport of water, the force of injection being supplied by the active cortex.

CHAPTER XX

MILKING THE PALM-TREE

LARGE quantities of sugar are obtained in India from the sap of Palm-trees. The Indian Date Palm (*Phoenix sylvestris*) grows to a height of 30 to 40 feet. The sap is drawn in a special way by wounding the upper end of the trunk (fig. 84). The daily yield of sugar-containing sap from this tree is often as high as 19 litres per day, a litre being equal to about two pints. The sugary sap is drunk fresh, or is used for the manufacture of sugar; it is also-fermented for making intoxicating liquor.

The other tree from which sugary sap is collected is the Palmyra Palm (*Borassus flabellifer*); it is a tree of very slow growth and lives for more than a century. The sap is drawn from the cut end of the flowering spike or spadix, which appears only in early summer. Fermentation soon becomes pronounced at high temperatures, and very special precautions have to be taken to obtain the sap unfermented. Ordinary cleaning of the collecting pot is found insufficient for the purpose; the prevailing custom is to smear the vessel with quick-lime and wash it afterwards. This antiseptic treatment is usually successful in securing the fresh sap as a drink.

The total quantity of sap given out by a single tree during its life may be as much as 120,000 litres. The sap is rich in sugar, the content being as high as 10 per cent. It may be regarded as a very efficient plant for the manufacture of sugar, since the total yield from a single tree during its lifetime may reach the enormous total of 12,000 kilograms or nearly 12 tons.

HOURLY VARIATION IN THE YIELD OF SAP

The rate at which the sugar-containing sap is given out is not the same at all hours of the day and night, but shows a very interesting variation. For the investigation of this, I devised the Tilter, by which definite quantities of sap collected are automatically recorded for hours and days. The recording apparatus may be placed inside the laboratory

FIG. 84. The Indian Date Palm.
The stem is sliced for collection of sap in the pot.

at any distance from the tree. As soon as the collecting cup of the Tilter becomes full, it overturns and empties itself. This upset completes an electric circuit each time, and the electric signal thus sent makes a mark on a recording drum. All that is necessary is to count the number of marks, which are closer together when the exudation is copious, and wider apart when the yield is scanty. The automatic records show that the rate of the yield of sap is at its minimum at about 1 P.M. and at its maximum at about 2 A.M. (fig. 85). The exudation at night is thus seen to be very much greater than in the day-time.

FIG. 85. Record of Exudation from the Indian Date Palm for 24 hours.

Note the separation of the dots after 1 P.M. and their closeness after 3 P.M.

EXPLANATION OF THE GREATER EXUDATION AT NIGHT

Careful measurement of the yield of a particular Date Palm showed that it gave 700 cubic centimetres of sap between six in the morning and six in the evening, while for the succeeding twelve hours of night the exudation was 2150 cubic centimetres, an amount three times as great. The explanation of this difference is to be found in the fact that the tree loses water by transpiration from the leaves as well as by the exudation of sap from the wounded surface. Consequently maximum transpiration should correspond with minimum exudation, and *vice versa*. As the loss by transpiration is feeble at night, the exudation of sap from the cut surface is relatively greater. During the day-time, on the other hand, the loss due to transpiration from leaves is very great ; hence the smaller yield of sap.

NECESSITY OF REPEATED WOUNDING

I now turn to the explanation of the exudation of sap by the Indian Date Palm. Many trees before the unfolding of leaves in early spring are filled with sap under considerable pressure, on account of which the sap exudes as soon as a hole is drilled into the tree. But in the Palm no sap exudes from a hole drilled in the trunk. I felled a tree, and the cut ends of the trunk did not exude a single drop of sap. Portions of the tissue taken from the interior of the trunk were found to be almost dry, and it was only after considerable compression that a small quantity of sap could be squeezed out. This experiment proves that there is no root-pressure to cause exudation from an injured surface. It should be remembered, in this connection, that the Date Palm grows in a dry or even arid soil; hence necessity compels the tree to exploit fully the scanty and precarious supply of water. The tree sends out more than a thousand roots to a considerable distance, each as thick as one's little finger. I followed each of these roots to a distance of more than twenty feet, yet the end was not in sight. The trunk of the tree is thus slowly charged with water absorbed by the enormously extended root-system, the sap being held in the trunk of the tree with great tenacity.

A hole made in the trunk, as stated before, does not yield a single drop of sap from the Palm. How, then, is the tree induced to give out its hoarded treasure? The normal inactivity of the wounded surface of the Palm is shown by the fact that there is no exudation at first even when vertical slices are cut in the upper part of the stem. The sap begins to exude only after repeated slicing of the stem for nearly a week.

What is the explanation of these facts? It has already been shown in a previous chapter that a living tissue may be roused from a state of inactivity to multiple activity by adequate stimulation. A very inactive tissue would naturally require a very strong stimulus or a succession of stimuli which became effective by their cumulative effects. In order to induce exudation from the Palm, the intense wound-stimulus had, therefore, to be repeated for several days in succession.

Process of 'Milking'

In the Palmyra Palm, in which the spike of flowers or spadix yields the sap, the process of coercion is somewhat different. There is no exudation when the tip of the spadix is cut off; exudation takes place only after the inflorescence has been subjected to a special treatment for several days. For this purpose two different processes have been elaborated in different countries, which may aptly be described as 'butting' and 'milking,' from the not very far-fetched analogy of the action of the calf to make the cow yield her milk.

The Malaya people strike the spike of the flower with a wooden mallet repeatedly for nearly a fortnight, after which the sugary juice is yielded from an incision. In India the method employed is perhaps more humane. The long spadix is held tightly between the fingers and kneaded from above downwards, the process being similar to the milking of a cow. This milking process is repeated day after day for a week. Cutting of the tip is then followed by a copious yield of sap. The preliminary hammering may be compared to the 'butting' of the calf, the 'kneading' to the usual process of milking.

The methods employed to induce the yield of sap by the previously inactive tissue of the Palm are fundamentally similar. They have one object in common—namely, the arousing of the dormant activity by the repeated application of stimulus, which may take the form of repeated cuts, repeated blows, or repeated kneading. As a result of this treatment the inactive tissue becomes as active as a glandular tissue, and is thus able to maintain the exudation even though there be no internal sap-pressure to urge it.

CHAPTER XXI

ACTION OF ALKALOIDS AND COBRA-VENOM ON PULSE-BEAT OF ANIMAL AND PLANT

THE electric method enabled me, as explained in a previous chapter, to prove that a propulsive tissue runs throughout the length of the plant, and that the flow of sap is brought about by a mechanism essentially similar to that of the circulation of the blood in the animal. The rhythmic pulsation of this tissue in the plant has, as already explained, been observed only by means of the corresponding electric pulsations.

The channel for the transport of sap may be regarded as an artery, the throbbing activity of which is its pulse-beat. Is it at all possible to record the actual pulse-beat mechanically? If we can succeed in doing this, the pulsatory activity of the plant will have been demonstrated by two entirely independent methods, the electrical and the mechanical.

There are yet other rigorous tests which may serve to demonstrate the essential similarity of the mechanisms for the circulation of the sap and of the blood. Various alkaloids induce characteristic reactions in the pulse-beat of the animal; do they similarly affect the pulse-beat of the plant?

RECORD OF THE PULSE-BEAT OF THE ANIMAL

The pulsations of the animal heart may be recorded directly by the Cardiograph, or indirectly by the Sphygmograph. The Cardiograph is essentially a magnifying lever, the short arm of which is attached to the beating heart, the longer arm recording the beats on a moving plate of smoked glass. Under a stimulating drug the pumping activity is greatly enhanced, the record showing an increased frequency or increased amplitude of pulsation. A depressing agent, on the other hand, brings about a diminution in the frequency or amplitude of the heart-beat.

The heart-beat has a corresponding pulse-beat in the arteries. The enhanced or depressed activity of the heart and the consequent changes of blood-pressure can be made out from the records given by the arterial pulse. In the human subject the radial artery is on the surface, at the wrist. The pressure-variation in the artery can be recorded by the Sphygmograph, which consists of a series of magnifying levers actuated by the pulse. This would obviously be impossible were the artery buried under other indifferent tissues instead of being on the surface.

The Plant-Feeler

It has been shown that the sap is propelled by the rhythmic pulsation of the propulsive layer in the stem. But any attempt to feel the pulse of the plant would, by the very nature of the case, appear to be hopeless, for the amount of expansion and contraction at each pulse would be beyond even the highest powers of the microscope to detect, since the amplitude would be something like a millionth of an inch. The active cells, moreover, are buried in the interior of the plant; how can the invisible and intangible be made accessible?

Let us follow the course of the sap-stream. As the plant is pumping the sap along the stem, the passage of each pulse must be attended by an infinitesimal expansion. After the brief passage of the pulse-wave, the stem will revert to its original diameter, until the next pulse reinflates it once more. The difficulty of recording the variation of the sap-pressure in the stem arises from the fact that the conducting channel, unlike the radial artery, is buried under other tissues; the area of the channel is, moreover, insignificant compared with the total section of the stem. The infinitesimal dilatation and contraction due to the pulse-wave can only be detected by a Plant-Feeler of surpassing delicacy and sensitiveness.

The Optical Sphygmograph

A supersensitive apparatus for the purpose has been realised in the Optical Sphygmograph, in which a compound magnification is produced by a lever and a rotating mirror. The stem is placed between two rods, one of which is fixed and the other movable. Two v-shaped pieces of ivory carried by the two rods touch two diametrically opposite points of the stem at R and *c*. The fulcrum-rod F of the movable lever is supported on jewel bearings. This lever is made of a porcupine quill, which combines lightness and

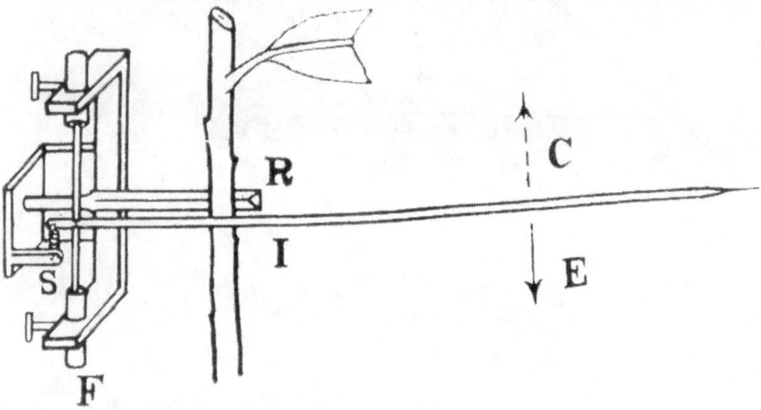

FIG. 86. Diagram of the Plant Sphygmograph.

rigidity in an unusual degree; its inertia being negligible, it quickly follows the movement of the pulse-beat. The lateral pressure of the lever on the stem is adjusted by the spring S, which may be placed either behind or in front of the lever. The length of the lever produces a magnification of about thirty times (fig. 86), which, however, is too slight to render the pulsations visible. It is therefore necessary to superpose a further optical magnification. A thin silk thread tied to the tip of the sphygmograph-lever makes one turn round a thin vertical rod supported both above and below by jewel bearings. The other end of the string is attached to a fine spiral spring, by which the pressure of contact on the stem can be easily adjusted. The outward movement of the lever L caused by the pulse-wave during expansion rotates the vertical rod in a clockwise direction, contraction producing anti-clockwise rotation. The rotation thus produced is highly magnified by a small mirror M, by which a beam of light is reflected on to a distant screen (fig. 87).

The total magnification by the Optical Sphygmograph is about a million times, which is sufficient for public demonstration of the characteristic effects of different alkaloids on the propulsive tissue of the plant. The complete apparatus is shown in fig. 88. The special applicator

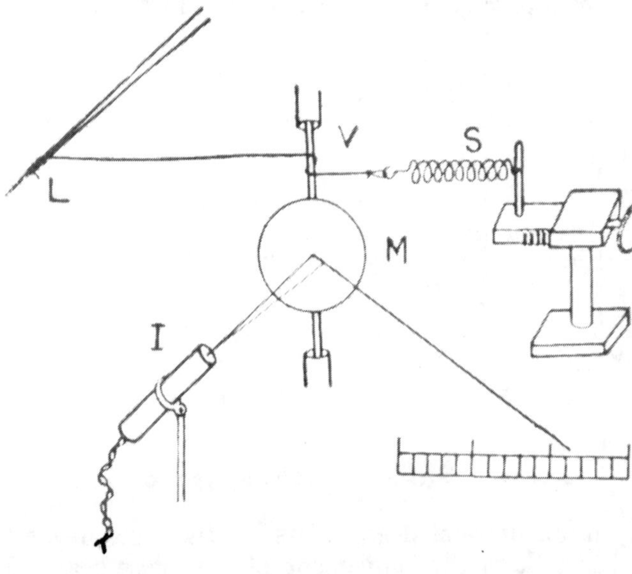

FIG. 87. Diagram of the Sphygmograph with optical attachment.

carries three small cups, containing respectively a stimulating, a depressing, and a poisonous solution. One or other of these may be rapidly applied to the cut end of the stem.

For the record of the individual pulse-beat of the plant a still higher magnification—namely, of ten million times—is necessary. This has been rendered practicable by my Magnetic Sphygmograph.

In the normal condition of the plant the average pressure remains constant, as shown by a practically horizontal record in the sphygmogram. This state of balance is not static but dynamic, consisting of oscillations about a position of equilibrium. The alternate expansion and contraction due to rhythmic activity are exhibited in the pulse-records,

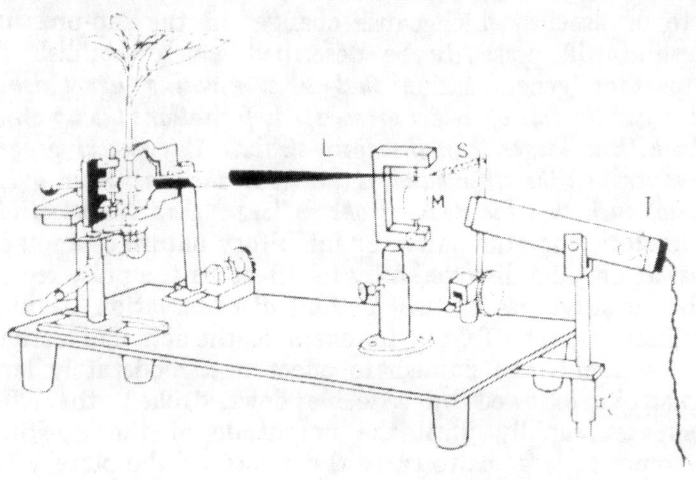

FIG. 88. The complete Optical Sphygmograph.

the dilatation by an up-curve and the contraction by a down-curve, these two being equal (fig. 89). As these pulsations are very minute, as already stated, it requires the very high magnification of ten million times for their demonstration. They are, however, conspicuously exhibited when the pressure is undergoing any change.

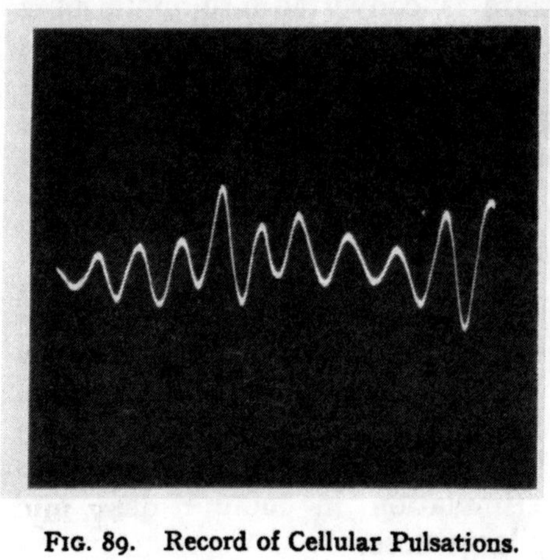

FIG. 89. Record of Cellular Pulsations.

PULSE-RECORDS UNDER STIMULANTS AND DEPRESSANTS

The blood-pressure in the animal, as previously explained, can be increased by a stimulant and lowered by a depressant. In the parallel experiment with the plant, the unique opportunity is presented of actually observing the action of the cellular pump, and the unequal strokes delivered by it during enhancement or depression of the rate of ascent, which cause changes in the sap-pressure. Results will presently be described which establish the important generalisation *that a stimulating agent which enhances the rate of ascent gives rise to pulsations the up-stroke of which is larger than the down-stroke. A depressing agent produces, on the other hand, a change in the constituent pulsations, such that the down-stroke is larger than the up-stroke.* Therefore, the stimulative or inhibitory nature of an agent can at once be discovered by its effect on the pulse-record. The sensitiveness of this method of observation is extraordinarily great. Taking, for example, the action of a stimulative agent, the immediate effect is a moderately large up-stroke followed by a feeble down-stroke; the effect increases rapidly until the amplitude of the up-stroke becomes so large as to carry the record off the plate. The frequency of the pulsations is also increased, so that the individual pulsations merge one into the other. The application of a depressant now produces the reverse change. It is interesting at this crisis to note the change in character of the individual pulsations, by which the hitherto increasing pressure is converted into diminishing pressure. After a period of hesitation, the down-stroke becomes predominant; a series of such constituent pulsations produce the descending curve which indicates the induced diminution of pressure.

Characteristic Action of Alkaloids

Alkaloids produce characteristic effects on the heart-beat of the animal. For the sake of convenience, they may be divided into three classes: (1) *stimulants*, which produce in general an enhancement of activity; (2) *depressants*, causing an inhibition of activity; (3) *stimulant-depressants*, producing stimulation in minute dose and depression in moderately large dose.

For the detection of changes of activity induced in the animal heart, the Cardiograph is more direct and sensitive than the Sphygmograph. But the only apparatus which records the pulse-beat of the plant is the Magnetic Sphygmograph. The similarity of effects of certain alkaloids on animal and plant will be demonstrated as follows:

Effect of	Animal Cardiogram	Plant Sphygmogram
Stimulant	Enhanced frequency	Increased sap-pressure. Up-stroke of pulse larger than down-stroke
Depressant	Diminished frequency	Diminished sap-pressure. Down-stroke larger than up-stroke

The result in both animal and plant is modified (1) by the tonic condition of the tissue and (2) by the dose and duration of application. I shall presently describe certain simple and typical examples.

The cardiogram was obtained with the heart of a fish, *Ophiocephalus*, in which the normal beat remains uniform for a great length of time. The sphygmogram of the plant was obtained with *Cosmos*, *Centaurea* and *Antirrhinum*.

Effect of Stimulants

Dilute solution of camphor enhances the activity of the animal heart. The stimulative effect is more easily shown with a specimen which is in a somewhat depressed condition. It is obvious that little further enhancement can be induced in a heart that is already in a condition of maximum activity. The normal record of pulsation of the animal heart is shown in the upper record of fig. 90. Injection of camphor, two parts in a thousand, considerably quickened the heart-beat, as seen in the lower record.

The effect of camphor on the plant is shown in fig. 91. The original amplitude of pulsation was feeble, so the record appears as a horizontal line. Application of dilute camphor produced a sudden increase of pressure shown by the ascending curve, in which the constituent pulsations show the up-stroke to be larger than the down-stroke. I obtained similar results with other stimulants, such as musk, provided the dose is not excessive.

ACTION OF DEPRESSANTS

Morphine may be taken as a typical depressant. The left record in fig. 92 shows the normal heart-beat of the fish; the record to the right shows the effect induced by

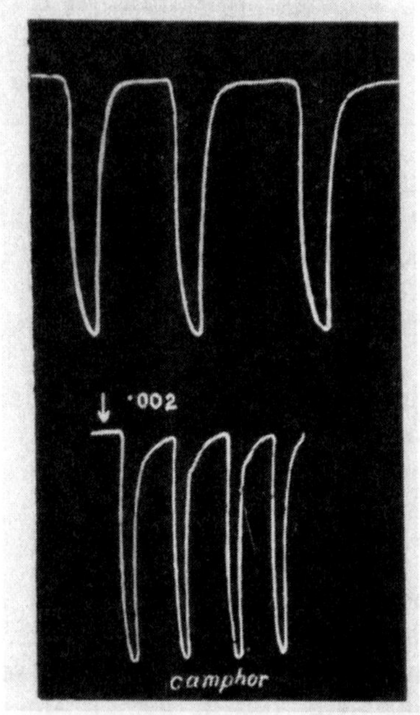

Figure 90.

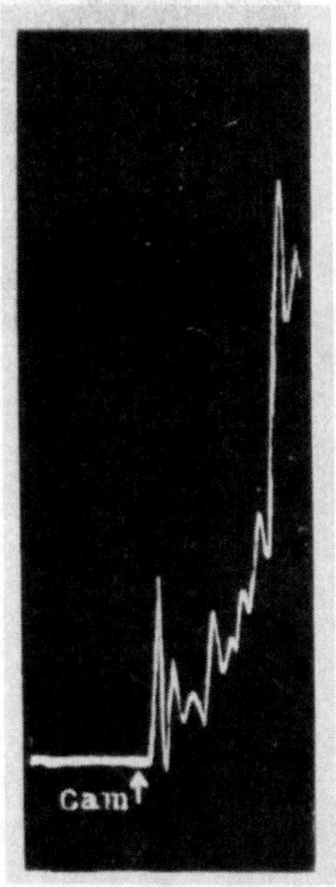

Figure 91.

FIG. 90. Effect of Camphor. Upper record shows normal rate of heart-beat. Rapidly descending line represents expulsion of blood by systolic contraction; ascending line represents active diastole; horizontal or slightly ascending line, passive diastole and pre-systole, the duration of which is greatly modified by action of drugs. Lower record represents stimulating action of Camphor in quickening the heart-beat. (Heart of Fish.)

FIG. 91. Effect of Camphor on the plant; enhancement of sap-pressure due to increased pumping activity. Note up-stroke of each pulse larger than down-stroke.

morphine, the amplitude and the frequency of pulsation exhibiting a depression.

The depressing action of morphine on the pulse-beat of the plant is seen in the next record, the constituent pulsations of the descending curve exhibiting down-strokes of each pulse larger than the upstroke (fig. 93). Alcohol, even in small doses, produces a depression of the beat.

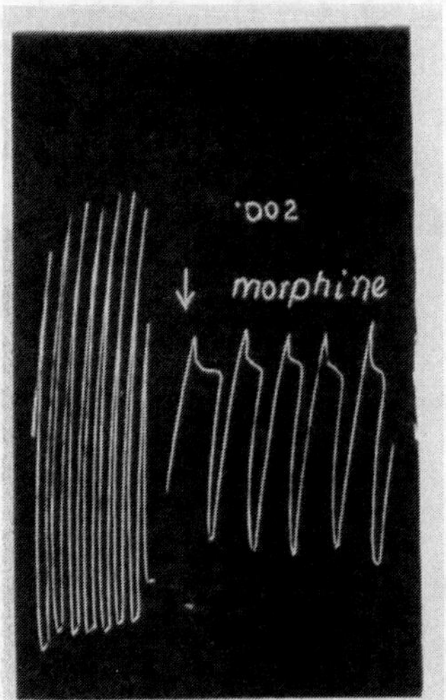

Figure 92.

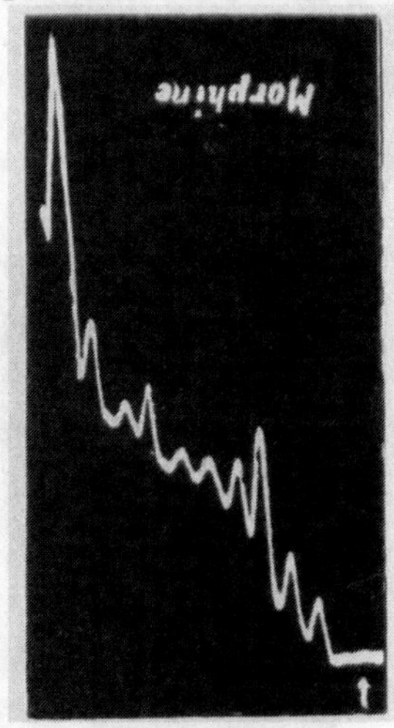

Figure 93.

FIG. 92. Action of Morphine: depression of cardiac activity in the fish.

FIG. 93. Diminution of sap-pressure in the plant.

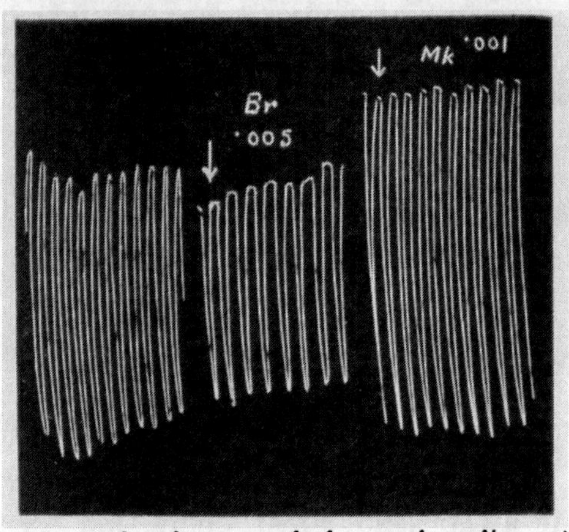

FIG. 94. First series gives record of normal cardiac activity; the second shows depression under KBr; and the third, enhanced activity under musk. (Fish.)

ANTAGONISTIC REACTIONS

I will next describe the remarkable records obtained by alternate application of a depressant and a stimulant to both animal and plant. For a mild depressant, I used a solution of bromide of potassium. In the first of the series of records in fig. 94 is shown the normal heart-beat. After application of bromide of potassium solution,

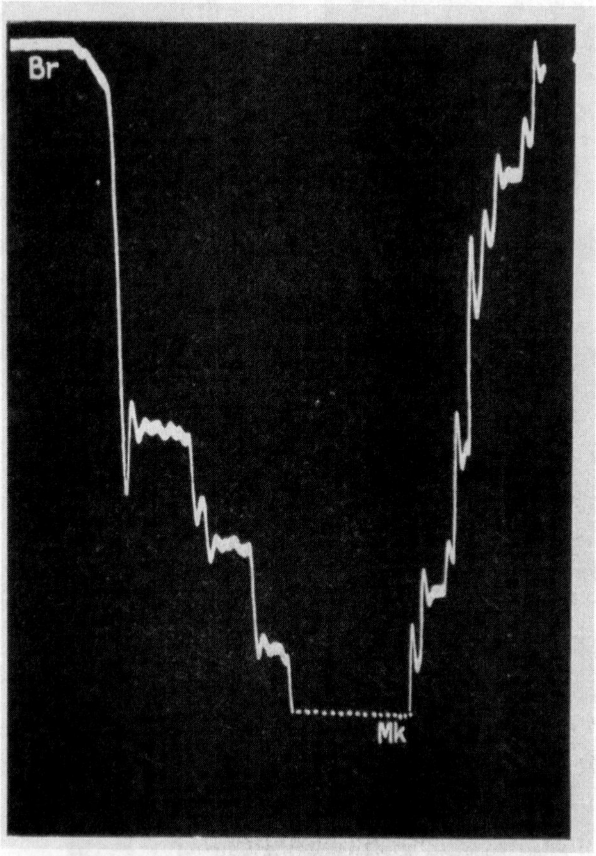

FIG. 95. Effect of KBr in depressing and of musk in enhancing sap-pressure.

five parts in a thousand, great depression occurred, both in amplitude and in frequency. A solution of musk, one part in a thousand, was next applied; it not only neutralised the depressing action of the bromide, but enhanced the activity above the normal. In the parallel case of the plant, the bromide was seen to produce great depression; subsequent application of musk not merely removed the depression, but caused a great enhancement of pumping activity and resulting increase of pressure (fig. 95).

Such antagonistic reactions are even more strikingly exhibited under the action of a poison and its antidote. Thus while under the continued action of morphine the plant was on the point of death, as shown by movement of the light-index to the extreme left, application of atropine revived the pulsation; the beam of light moved to the right, indicating the revived life of the plant.

STIMULANT-DEPRESSANTS

A dose of strychnine, one part in a thousand, caused a marked enhancement of the cardiac activity, as seen in the

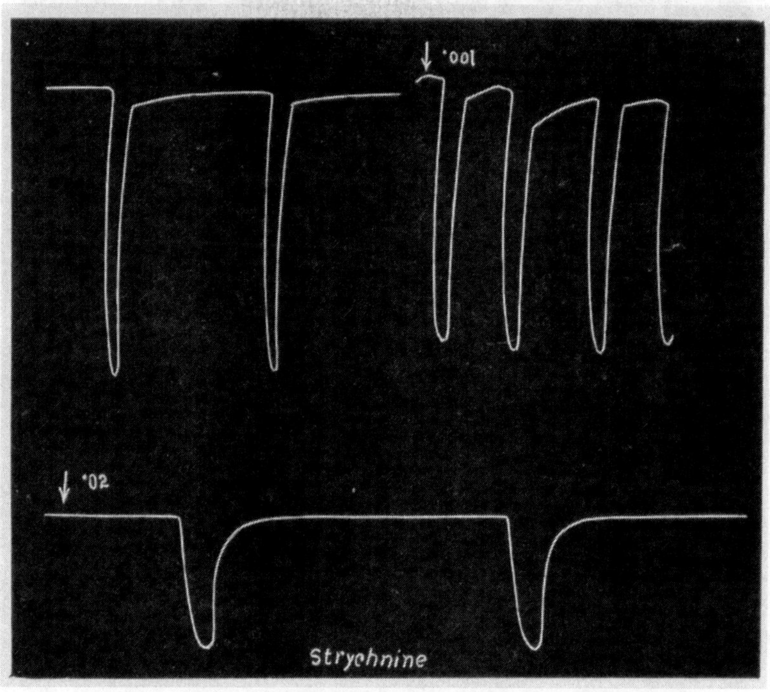

FIG. 96. Action of Strychnine. Increased frequency of cardiac pulsations under 0·1 per cent. solution (upper record). Depression and arrest under 2 per cent. solution (lower record).

quickened rate of pulsation; a 2 per cent. solution produced, on the other hand, a great depression and ultimate arrest of the heart-beat (fig. 96).

Parallel effects of stimulation and depression were produced in the plant by minute and large doses of strychnine. A dose of a solution of one part in a thousand acted as a stimulant, increased the sap-pressure: but one of a 1 per cent. solution caused marked depression and diminution (fig. 97).

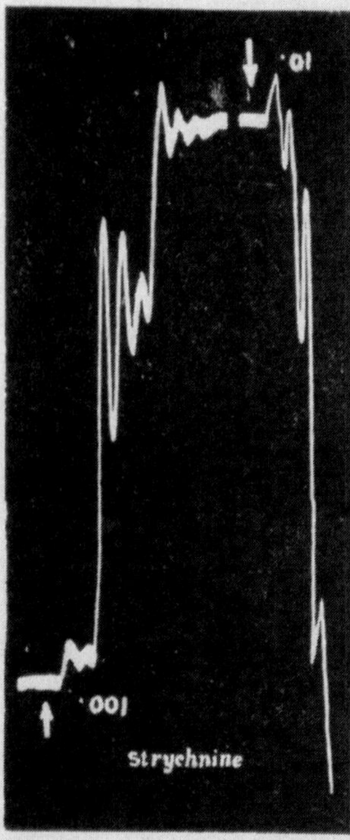

FIG. 97. Action of Strychnine. Enhancement of sap-pressure under a minute, and diminution under a strong, dose.

ACTION OF COBRA-VENOM

Cobra-venom acts on the animal as a very deadly poison. When injected subcutaneously it proves fatal when the quantity is as small as 0.00002 gram. There is a popular belief in India that even when a person bitten by the snake exhibits all the signs of death, such as stoppage of respiration and of pulse-beat, yet there may be a chance of revival. The dead body is, therefore, not cremated as in other cases of death, but placed on a raft and allowed to float down the river.

I investigated the effect of cobra-venom on both animal and plant, and employed for my experiments desiccated venom, which keeps its deadly property unchanged for many years.

EFFECT OF MODERATE DOSE OF VENOM

After taking the normal record of the vigorously pulsating heart of the fish, 0.5 c.c. of solution of venom, one part in a thousand, was injected into the vein. The amplitude of the pulsation exhibited a rapid decline and the heart-beat came to a complete stop in the course of twelve minutes. Shortly before this, a muscular spasm occurred in the body of the fish, shown as a down-twitch in the record (fig. 98).

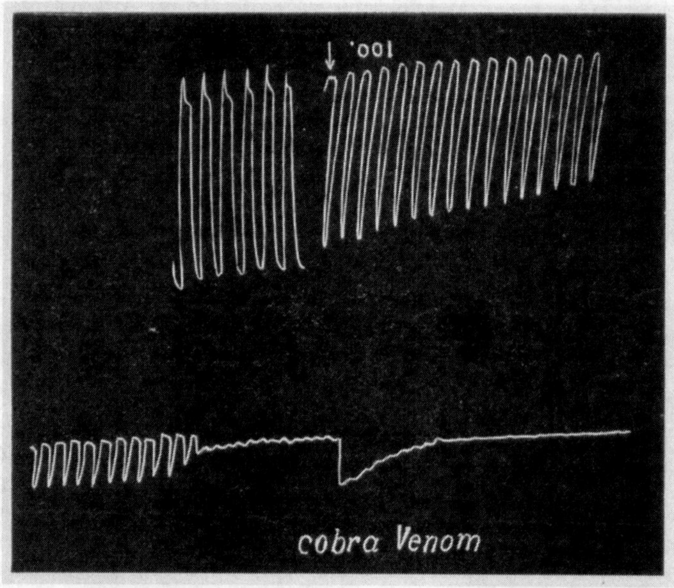

FIG. 98. Effect of Cobra-venom: Normal cardiac pulsation; rapid diminution of pulsation and final abolition after application of 0·1 per cent. dose of venom, as seen in lower record. Note death-spasm. (Fish.)

A solution of 1 per cent. of the venom was next applied to the plant. This produced first a great depression, as seen in the down-record, and in a short time the pulse-beat came to a permanent stop (fig. 99). Subsequent wilting and decomposition showed that the plant had been killed.

EFFECT OF *Shuchikavaran* ON CARDIAC PULSATION

The effect of a minute dose of cobra-venom, one part in a hundred thousand, produced in the plant an acceleration of activity and an enhanced rate of ascent of sap exhibited by increased pressure (fig. 100). The cut stem was placed in this solution for several days, and the specimen was found to remain in a vigorous living condition.

This led me to investigate the effect of a minute dose of cobra-venom on the cardiac activity. In this connection, I was greatly interested to find that a preparation known as *Shuchikavaran*, the principal constituent of which is a minute quantity of cobra-poison, has been employed in the Hindu System of Medicine for nearly a thousand years.

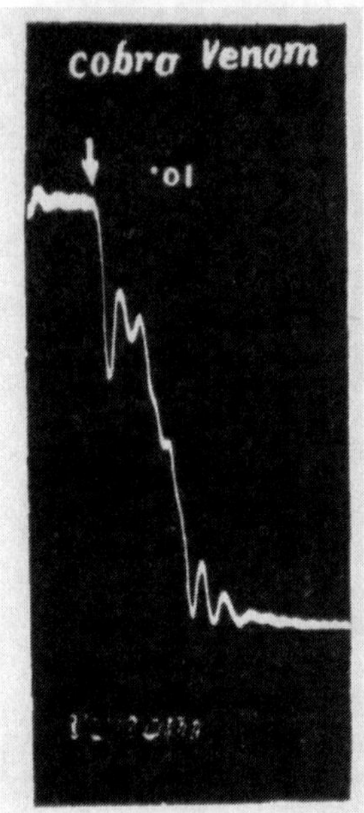

Figure 99.

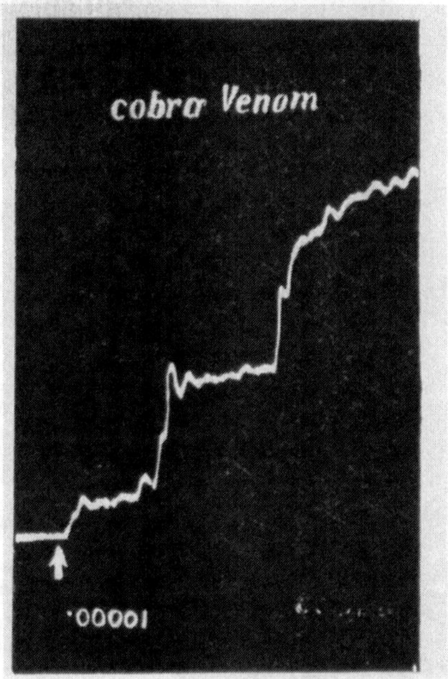

Figure 100.

FIG. 99. Effect of cobra-venom in diminishing sap-pressure and abolition of activity.

FIG. 100. Effect of highly dilute cobra-venom in enhancing sap-pressure.

Its use coincided with the rise of a new school of investigators who made a systematic study of the medicinal properties of various alkaloids and metallic compounds. *Shuchikavaran* is still used in desperate emergencies when the patient, in the last stage of illness, is on the point of death from heart-failure. It is said that in such emergencies this preparation of cobra-venom is most effective in reviving and strengthening the cardiac activity.

In order to observe the effect of *Shuchikavaran* on the

animal heart in a depressed condition, I injected a dilute solution of the preparation into the sinus of the fish. This

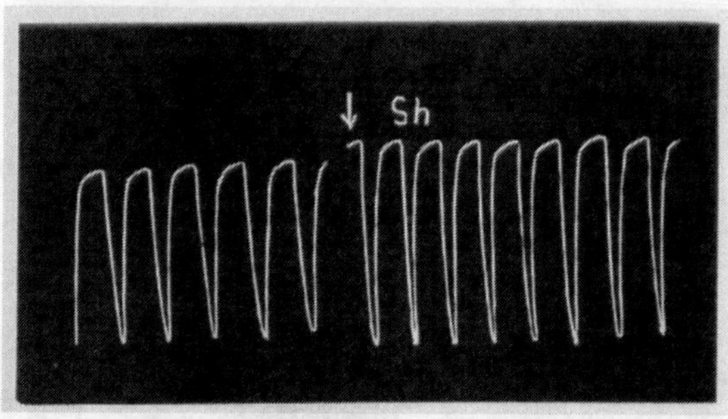

FIG. 101. Effect of ***Shuchikavaran*** in stimulating the heart.

produced a decided improvement in the frequency and amplitude of the pulsation (fig. 101). In other cases irregular pulsations became quite regular after the injection.

CHAPTER XXII

CARBON-ASSIMILATION

THE incessant activities of life require expenditure of energy that has been previously stored by the organism. Taking, for example, the rise of sap, the ceaseless pumping activity of the propulsive tissue raises enormous quantities of water to a considerable height. The energy of doing this work issues from the breakdown of organic chemical substances in internal combustion or respiration. The loss of energy must be restored by absorption and storage of energy from outside.

The supply of energy may be maintained in either of two ways: it may be absorbed by the organism in the active or kinetic form; or in the inactive, potential or latent form. The former is the method characteristic of plants; the latter of animals. The plant, in virtue of its green colouring matter, chlorophyll, can absorb the kinetic energy of the sun's rays and is thereby enabled to build up from the carbonic acid gas of the atmosphere, organic matter, that is, complex carbon compounds in which is stored, in the latent form of chemical combination, the energy of the absorbed light. The organic matter so produced is partly used in the maintenance and growth of the plant-body and consumed in carrying on its various vital activities. The remainder is stored in the tissues of the plant as reserve-material for future growth and, more particularly, in seeds and fruits for reproduction.

The animal, on the contrary, is entirely dependent for its supply of both energy and material upon organic food, which has been, originally at least, produced by green plants. Thus the animal is dependent upon the plant; and both are dependent upon the sun, the plant directly, the animal indirectly. The sun is, in fact, the prime source of all the energy that is set free in the living organism as heat, electric current, or movement; and, more generally, in the processes of combustion. To stand before a coal-fire is to bask in the sun which shone millions of years ago in the Carboniferous Period.

Hence the carbon-assimilation of the green plant, Photosynthesis, as it is now termed, is a process of the greatest theoretical and practical interest which deserves thorough investigation with a view, more particularly, to determine the conditions that affect its activity. This can be effected by accurately measuring the gaseous interchange between the plant and the atmosphere during photosynthesis, that is, either the volume of carbon dioxide absorbed, or the equal volume of oxygen evolved. The measurement of the intake of CO_2 involves complicated chemical analysis and is too long and laborious a method. The measurement of the output of oxygen is more promising. I have, in fact, been able to devise an apparatus on this principle, which I will now describe.

Automatic Record of Assimilation

Water-plants obtain their carbon from the carbonic acid dissolved in the water. When sunlight falls on these plants, carbonic acid gas is broken up, the carbon becomes fixed in the form of organic compounds known as carbohydrates, and an equal volume of oxygen is evolved which rises as a stream of bubbles from the plant. The rate of evolution of oxygen indicates the rate of assimilation.

Numerous difficulties were encountered in making this method practical; they have been overcome by my Automatic Recorder. A piece of a water-plant, *e.g. Hydrilla verticillata*, is placed in a bottle completely filled with tank-water containing sufficient CO_2 in solution, the open end of which is closed by a special bubbling-apparatus, the Bubbler, for measuring the oxygen evolved. The Bubbler consists of a U-tube, the further end of which is closed by a drop of mercury acting as a valve. The oxygen evolved by the plant, entering the U-tube, produces an increasing pressure, which eventually lifts the mercury valve and allows the escape of a bubble of the gas. The valve then immediately closes until it is lifted once more for the escape of another equal volume of gas. The movement of the mercury completes an electric circuit, which either rings a bell or makes an electro-magnetic writer inscribe successive dots on a revolving drum (fig. 102). The automatic method eliminates all personal errors of observation; it is so extremely sensitive that it is possible to measure a deposit of carbohydrate as minute as a millionth of a gram.

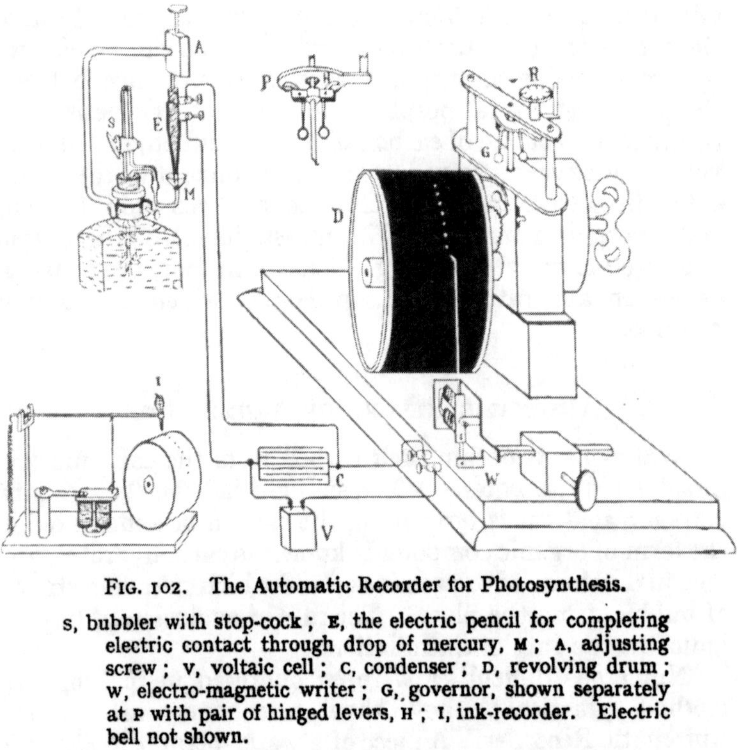

FIG. 102. The Automatic Recorder for Photosynthesis.

S, bubbler with stop-cock; E, the electric pencil for completing electric contact through drop of mercury, M; A, adjusting screw; V, voltaic cell; C, condenser; D, revolving drum; W, electro-magnetic writer; G, governor, shown separately at P with pair of hinged levers, H; I, ink-recorder. Electric bell not shown.

In illustration of the practical working of the apparatus I will give the following example. The plant with the apparatus is so placed as to face the northern light; the bell rings each time it has evolved a certain amount of oxygen representing an equal volume of absorbed CO_2. If a person now stands obstructing the light, the assimilation is slowed down and the bell now strikes at longer intervals. When strong sunlight is thrown on the plant, the successive strokes on the bell become greatly quickened.

The plant is such a sensitive detector of light that it may be employed as a photometer for indicating the slightest variations in the intensity of the light of the sky.

I have been successful in devising a contrivance by which the plant actuates an electric switch and turns on a light as soon as the sky-light is dimmed by a passing fog, the switch being turned off as soon as the sky becomes clear. This device may prove useful during winter in London. Still more interesting is the automatic record given by the electro-magnetic writer in the form of successive dots inscribed on the drum. When the rate of assimilation is in any way enhanced, the dots come close together. Depressed assimilation is indicated by widely separated dots.

The Hourly Variation of Assimilation

At what hour of the day does the plant assimilate CO_2 at the quickest rate? For this determination I took successive records from 7.30 A.M. until 5 P.M. for five minutes at a time. When the sun rose at 6.45 A.M. the light was too feeble to be effective. At 7.30 A.M. assimilation began, and the plant evolved four bubbles of oxygen in the course of five minutes; with the progress of the day it became more and more hungry, until at 1 P.M. it took four times as much CO_2 as it took in the morning. It is curious that the plant should also be most hungry at our lunch-time! The real reason for increased assimilation at 1 P.M. is the favourable conditions of light and temperature. The activity declined in the afternoon and became arrested with the onset of darkness (fig. 103).

Effect of Stimulation on Assimilation

As assimilation of food is essentially a vital process, any irritation by excessive stimulation proves to be highly detrimental. Thus a plant which was assimilating actively and showed its enjoyment by ringing the bell with great rapidity became suddenly depressed after receiving a strong electric shock. It was 'off its feed' for a long time, as indicated by the stoppage of the tinkling of the bell. The stronger the stimulation the more prolonged the stoppage of assimilation. The obvious moral is that we should take

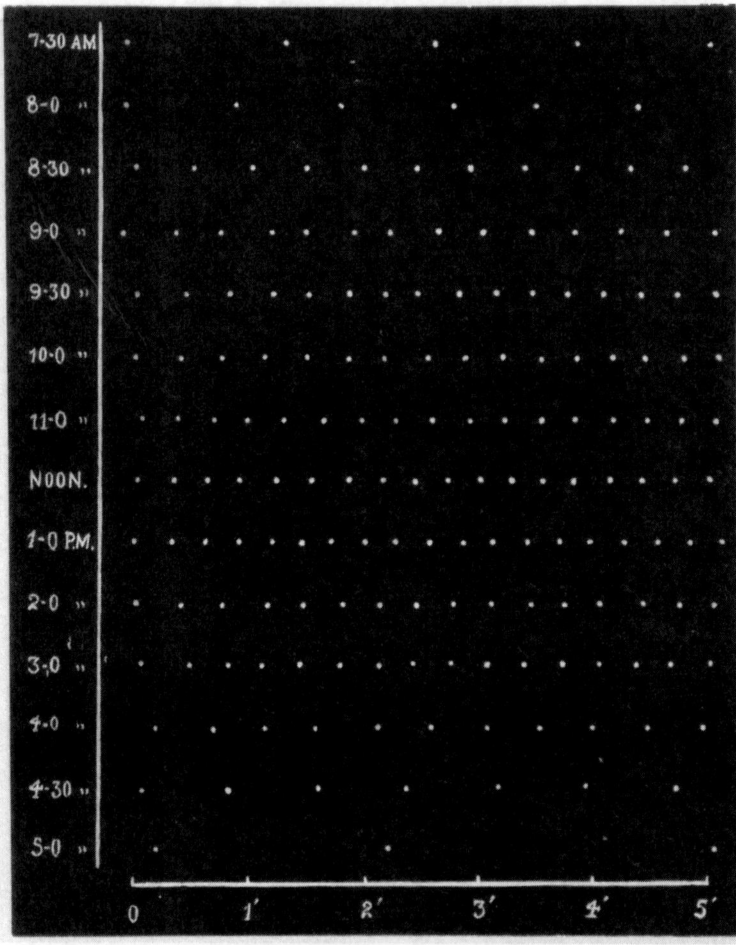

FIG. 103. Automatic Record of successive Bubblings for five minutes each, at different times of the day.

Note the slow rate at 7.30 A.M. and 4.30 P.M., and the quick rate at mid-day.

absolute rest for at least half an hour before meals. Food may indeed turn into poison when we are under an intense irritation.

Effect of Infinitesimal Traces of Chemical Substance

During the course of my investigations I discovered that the presence of the minutest traces of certain chemical substances induced an extraordinarily great enhancement of carbon-assimilation. The dilution employed was one part in a billion (billion in French measure is equal to 1000 millions). With certain substances a dilution of one part in two billions produced an increase of activity of more than a hundred per cent. The activity declined when the strength of the solution was raised above a critical dose. Dilute extract of thyroid gland, in a dilution of one part in a billion, produced a maximum increase of activity of about 70 per cent. The noticeable fact in the action of thyroid extract is that no diminution of activity below normal took place for a considerable range of dilution. The effect of traces of iodine was more or less similar. At first sight it is inconceivable that infinitesimal traces of certain chemical substances should have such a potent influence on life-activity. The immediate and concrete demonstration of the effect of minute traces of chemicals on carbon-assimilation is of special interest, since it enables us to understand the effect of ultra-measurable quantities of vitamins on general assimilation and of hormones on physiological reaction.

Efficiency of Green Plants in the Storage of Solar Energy

The economic life of the present age may be said to be dependent to a great extent on the utilisation of the solar energy that has been stored in past ages by vegetable life. What is the efficiency of the plant-mechanism for the storage? It has hitherto been regarded as extremely low, less than 1 per cent.: but the methods hitherto employed in this determination have been more or less defective. I therefore undertook a careful re-determination by new and highly sensitive methods. The incident solar energy was carefully determined by my Magnetic Radiometer, and the energy stored by the plant was also accurately measured. The efficiency was found to be much higher than had been generally supposed, being as high as 7.4 per cent. It is interesting to compare the efficiency of transformation in an ordinary steam-engine with that in the photosynthetic organ. In the former the potential energy of coal is transformed into the kinetic energy of motion; in the latter the kinetic energy of radiation is transformed into the potential energy of complex chemical compounds. The efficiency of the photosynthetic organ may be taken as about half that of an ordinary steam-engine. After all, it may not be such an unpractical proposition to devise a chlorophyll apparatus for trapping sunlight.

CHAPTER XXIII

THE NERVE OF PLANTS

WHEN the tip of one's finger is gently scratched, an impulse is created which is perceived in the brain as sensation. The message is transmitted along a nerve-thread, which is the definite channel for the conduction of impulse; when the nerve is injured in any way there is an end of all sensation. The scratch and the resulting sensation appear to be simultaneous, but in reality a short time is required for the impulse to travel from the finger-tip to the brain. The speed of the nervous impulse may be found somewhat as follows: the person on whom the experiment is made gives a signal when he feels the sensation of the scratch on his toe. The interval between the scratch and the signal enables us to calculate the speed of the impulse through the length of the nerve.

If the nerve terminates in a muscle, then the arrival of the impulse is signalled by the twitch of the muscle. Experiments on nervous impulse are usually carried out with a piece of nerve and muscle of the frog, which may be detached and kept alive for several hours. If now a distant point of the nerve be stimulated by an electric shock, an impulse is transmitted along the nerve to the terminal muscle which is attached to a recording lever. The record is taken on a moving drum, on which time-marks are inscribed by a chronograph. The speed of the impulse is found from the length of the nerve and the time recorded for transmission.

In the nervous circuit of the animal three different parts may be distinguished. The first is the 'receptor,' which receives the shock from outside; the second is the 'conductor,' the 'nerve,' by which excitation is carried to a distance, though no visible change occurs in this conducting tissue during the transmission of an impulse. Finally, the impulse impinges on the terminal responding organ, the 'effector,' which may be a muscle; the response is then visibly manifested by movement.

The beginnings of such a nerve and muscle system are seen in animals as low as the sea-anemone, where stimulation of its tentacles causes a motile reaction at a distant part, there being no movement in the intermediate region. The receptor and the effector are thus at a distance from each other, the connecting link being the nerve.

This mode of transmission of excitation, where the effect of a stimulus applied at a point is manifested by a movement at a distance, would appear to be not unlike what occurs in the Sensitive Plant *Mimosa pudica*. Here also the application of stimulus, say of an electric shock, to one of the subpetioles, gives rise to an impulse which, travelling onwards along the leaf-stalk, reaches its motile organ, the pulvinus, the contraction of which produces the sudden fall of the leaf. Though the effects produced in the plant and in the animal are so similar, yet the prevalent opinion has been that impulse travels in the plant in a manner quite different from that in the animal nerve.

Water-Pipe or Nerve?

Let us consider on what experimental fact this conclusion is based. Pfeffer gave a knife-stab to the plant to stimulate it, and observe how the plant answered under such brutal treatment. Imagine expecting a man to give any rational answer when the stimulus employed is a knife-thrust; so far from giving any coherent response, he would be thrown into convulsions! After the knife-thrust into the plant, Pfeffer noticed the escape of sap from the wound. The turgid stem of the plant he imagined to be like an india-rubber tube filled with water, the escape of sap being supposed to produce a sudden diminution of pressure, causing a traction or pull on the sensitive pulvinus. This hydro-mechanical impulse is considered practically equivalent to the movement of water in a pipe.

Ricca is also enamoured of the method of knife-thrust

in causing stimulation. He imagines that the wounding of the wood causes it to secrete some stimulating substance, a hormone, and that this hypothetical hormone is conveyed by the ascent of sap to the leaf, which it stimulates to movement. This is a misapplication of the theory of the hormone as enunciated by its authors, Starling and Bayliss, who insisted on the fundamental difference between the two modes of communication between two distant organs, by *transfer of matter*, and by *transmission of motion*. The first is exemplified by the *slow* movement of liquids carrying chemical stimulants in solution, such as occurs in the ascent of sap in the plant, or in the circulation of blood in the animal; the second is the *rapid* conduction of excitation from point to point associated with the propagation of nervous impulse. These two different modes have been aptly likened to communication by post or by telegraph. The difference between the two speeds is so great that it would be an unpardonable mistake to confuse one with the other.

The hydro-mechanical theory and the theory of hormone-transport by ascent of sap are both based on the supposition that a wound is necessary to produce a mechanical disturbance or the secretion of an irritant causing stimulation. They stand condemned if it can be shown that an excitatory impulse is generated and conducted in the plant by a feeble stimulus and without any wound.

The plant is highly excitable, and a very feeble stimulus is sufficient to start an impulse. The only excuse for using a knife-thrust as a stimulus is the erroneous supposition that plants are very much less sensitive than animals, and must therefore be goaded into activity by violence. This is a gratuitous and totally unfounded assumption, for, as previously stated, I discovered that Mimosa can be excited by an electric shock of one-tenth of the intensity of that which evokes human sensation. No wound is produced, yet the excitation is transmitted to a considerable distance. This result alone is sufficient to show the totally unfounded character of both the hydro-mechanical and hormone-ascent theories. I shall describe other crucial experiments which will completely disprove them.

THE THEORY OF SAP-MOVEMENT

A hypothetical stimulant is, according to this theory, conveyed to the leaf in the ascending sap. The impulse in this case should always travel upwards in the same direction as the sap, and not downwards against that direction. The speed of the impulse, moreover, should be the same as that of the sap-movement. I applied the

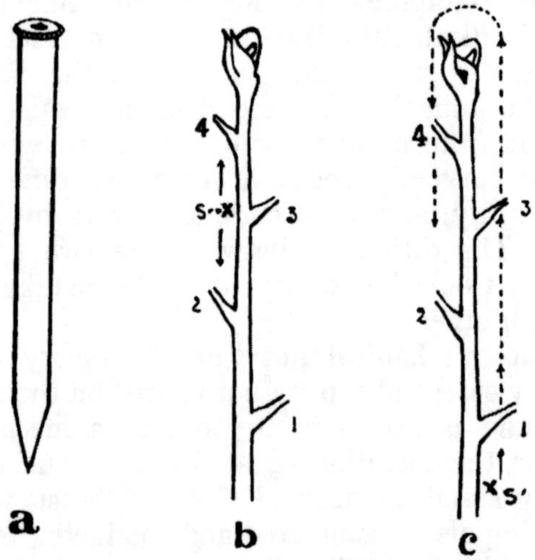

FIG. 104. Effect of Unilateral Scratch Stimulation.

a, the Scratch-Stimulator; b, effect of stimulation of moderate intensity, s, applied on left gives rise to simultaneous impulses up and down; c, effect of strong stimulus s′ applied on right side causes an impulse which ascends on the right side and then descends on the left.

stimulus of a superficial scratch on one side of the stem of Mimosa, and found that the excitation travelled simultaneously both upwards and downwards, causing the fall of leaves both above and below. On increasing the intensity of the stimulus, the excitation was found to ascend along one side of the stem, reach the apex, and then descend down the other side (fig. 104). Rising sap could not possibly have produced such characteristic results. The conduction of excitation up and down the stem can only be attributed to the presence of a special conducting tissue, that is a nervous tissue.

Accurate measurements which I have carried out show, moreover, that the rate of transmission of excitation is several hundred times quicker than that of the ascent of sap. That the movement of sap has nothing whatever to do with conduction of excitation finds a further striking demonstration in the following experiment: A drop of hydrochloric acid was applied to the tip of the uppermost leaf of Mimosa. The impulse generated travelled to a considerable distance downwards against the direction of the normal ascent of sap; subsequent chemical examination proved that the stimulant had not been transported, but had remained localised at the point of application.

Having proved that the transmission of excitation is neither hydro-mechanical nor due to the movement of sap, I shall next adduce evidence to show that the conduction in the plant is a propagation of protoplasmic excitation, just as in the excited nerve of the animal.

Tests for Nervous Impulse

It is obvious that the mechanical movement of water through a pipe will not be affected by heat or cold within reasonable limits; the pipe will not lose consciousness and stop the flow of water if it be anæsthetised, nor will its conducting power be abolished by applying round it a bandage soaked in poison. These agents will, on the other hand, profoundly affect the transmission of nervous excitation. The nature of an impulse, whether mechanical or nervous, may be discriminated by several crucial tests.

If physiological changes affect the rate of conduction, then the impulse must be of a nervous character; the absence of any such effect, on the other hand, proves the mechanical character of the impulse.

There are various physiological means by which it is possible to retard or arrest the nervous impulse, but have no such effect on a mechanical impulse. Some of these are as follows:

(1) When the conducting tissue or nerve is cooled, the speed of the impulse is slowed down, culminating in its arrest.

(2) Poisonous solutions applied on the nerve permanently abolish its conducting power.

(3) The conducting power is temporarily arrested by a block produced by the passage of an electric current in a portion of the nerve through which the impulse is being transmitted. This electrotonic block is removed on the stoppage of the current.

Measurement of Velocity of Impulse

The tests are carried out by automatically determining the velocity of impulse first under normal conditions, and then under other conditions which are known to modify the transmission of excitation in animal nerve.

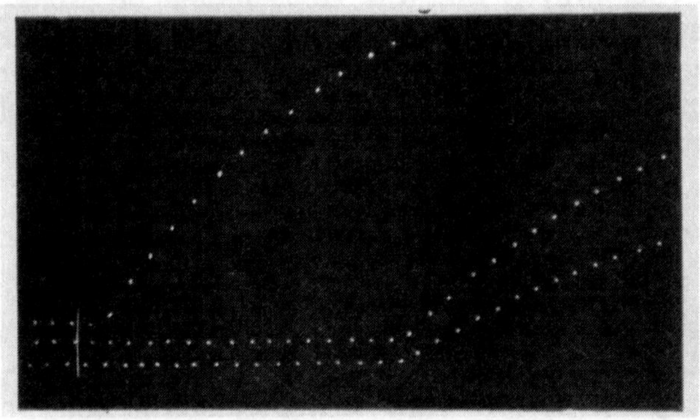

FIG. 105. Determination of velocity of transmission of excitation in the petiole of Mimosa.

Two lower records are in response to indirect stimulation applied at a distance of 30 mm.; upper record of response to direct stimulation gives the latent period. Recorder 10 v. per second.

The normal velocity is found by applying a definite intensity of electric shock on the leaf-stalk at a specified distance, say 30 mm., from the motile pulvinus, the intensity of stimulus being maintained constant in successive experiments. The moment of application of the shock is marked in the record by a vertical line. The Resonant Recorder taps successive dots at intervals, say of one-tenth of a second, and thus measures the interval between the application of stimulus and the resulting movement of the leaf indicated by an up-curve. The interval between stimulation and response is seen to be 16·2 spaces, each of the value of 0·1 second (fig. 105). The total interval is therefore

1·62 second. In order to prove the reliability of the method, two successive records are taken, which show that the time required under normal conditions is practically the same in both experiments. The recorded time includes the latent period of the pulvinus, which represents the time consumed in starting the motor mechanism into action. The length of the latent period, that is, the interval between stimulation and response, is determined by applying stimulus directly to the pulvinus. In the present case this was 0·12 second, and the actual time of transmission of the impulse through a distance of 30 mm. is therefore 1·62 − 0·12 second, or 1·5 second. The velocity of the impulse is therefore $\frac{30}{1 \cdot 5}$ or 20 mm. per second.

The velocity of transmission of excitation is appropriately modified according to the vital condition of the plant. It is greater in summer than in winter. Another curious fact observed is that while a stout specimen responds in a leisurely manner, a thin specimen attains its acme of excitation in an incredibly short time. Such a difference is not unknown even in the human species. In a thin leaf-stalk of Mimosa the speed may be as high as 400 mm. per second, or 24,000 mm. per minute. While the velocity of impulse in Mimosa is lower than in the higher animals, it is considerably greater than in the lower animals, such as Anodon. The velocity in the plant may therefore be regarded as halfway between the two. The velocity of transmission in both Mimosa and in animal nerve is increased, within limits, by rise of temperature and diminished by a fall. In Mimosa a rise of temperature of about 9° nearly doubles the velocity.

Arrest of Impulse by a Physiological Block

A certain length of the structure through which the impulse has to travel can be so treated as to retard or arrest its passage (fig. 106).

Block by cold.—When part of the petiole was moderately lowered in temperature by the application of cold water, the transmission-time is shown by the record to have been prolonged. Excessive cooling by ice-cold water abolished the power of conduction (fig. 107). In order to show that the local application of cold abolished conductivity without affecting the excitability of the pulvinus, a direct

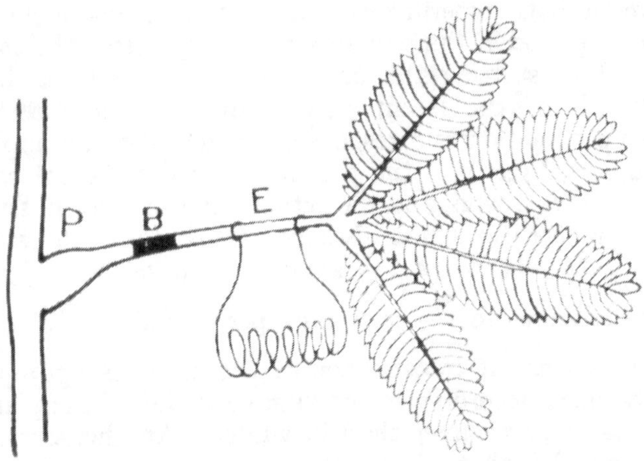

FIG. 106. The various physiological blocks are interposed at B, in the path of conduction.

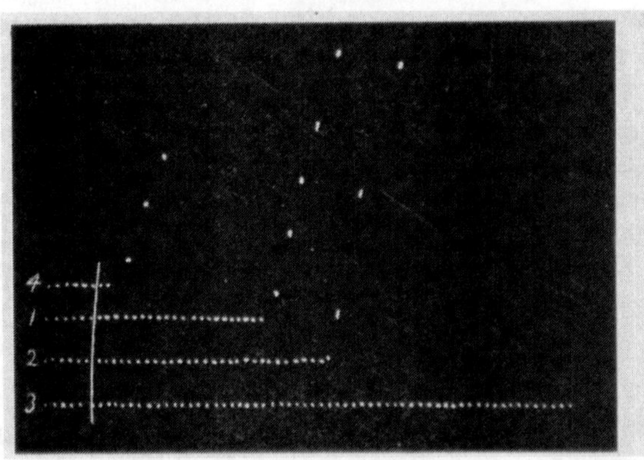

FIG. 107. Effect of Cold in inducing retardation and arrest of Transmission.

(1) Normal record; (2) Retardation due to slight cooling; (3) Arrest of conduction brought about by intense cold; (4) Record of direct stimulation.

shock was given to the pulvinus which exhibited its normal response.

In connection with this subject, I came across the

interesting phenomenon of paralysis of conduction as an after-effect of intense cold, this paralysis persisting for over an hour, even after the return of the tissue to normal temperature. Electrical shocks are found to be effective in curing certain forms of paralysis in human subjects, and I discovered the very suggestive fact that the lost power of conduction in the plant can be very quickly revived by subjecting the paralysed portion of the petiole to the action of tetanising electric shocks.

Block by Poison.—I will next demonstrate the action of poison in permanently abolishing the conducting power. A

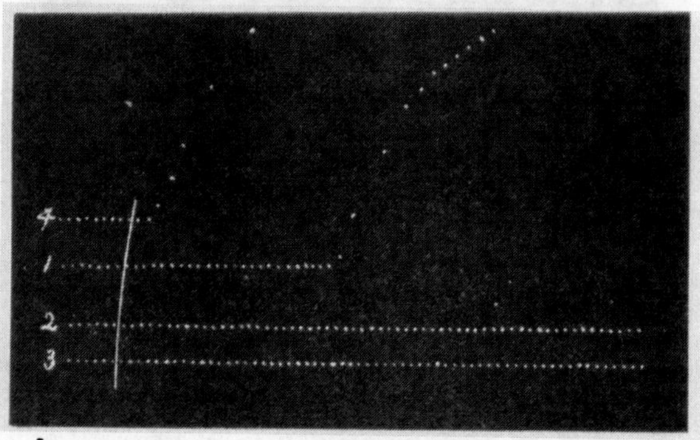

FIG. 108. Abolition of Conductivity by the action of Potassium Cyanide.

(1) Normal record; (2) Arrest of conduction after application for 5 minutes; (3) Record showing arrest of impulse, even with strong stimulation; (4) Record of direct stimulation.

strip of cloth half an inch in breadth was wound round the intervening length of the leaf-stalk, and a poisonous solution of potassium cyanide was applied. The effect of this was so great that the conducting power was abolished within as short a time as five minutes. Though the intensity of shock was increased some eight times, yet there was no response. Direct stimulation of the pulvinus showed that its motility had undergone no change (fig. 108).

Electrotonic Block.—In the above case the conductivity was permanently abolished; in animal nerve, as already stated, a temporary block is produced by maintaining a constant electric current in the path of conduction. The block persists so long as the electric current is maintained, and the conducting power is immediately restored on the stoppage of the blocking current. An exactly similar result was obtained with Mimosa; the block was put on at B and afterwards removed. This was repeated twice, and the record (fig. 109) shows that the transmission of excitation

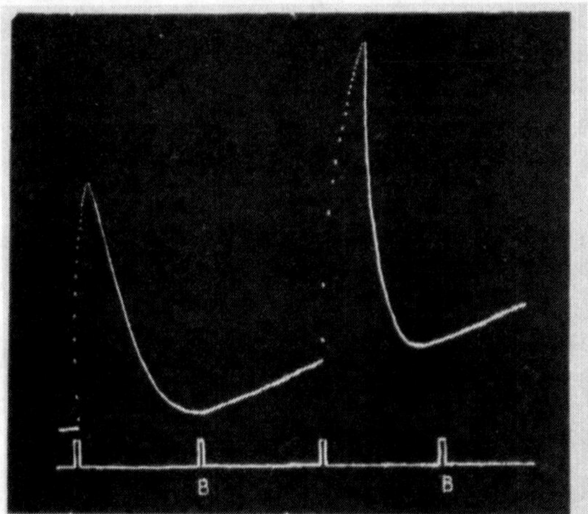

FIG. 109. Record of transmitted excitation with the block off and on. Arrest of transmitted excitation under electrotonic block B, B.

was invariably arrested whenever the blocking current was applied, and was restored on the cessation of the current.

POLAR EXCITATION BY ELECTRIC CURRENT

I will now adduce positive evidence in proof of the identical character of the transmission of excitation in the animal and in the plant.

It is well known that an electric current causes specific exciting effects in the animal nerve. On suddenly sending a current through a nerve, excitation is produced at the point of the nerve which is the kathode, that is, where the current leaves the tissue. On sudden stoppage of the current, excitation is transferred to the anode, where the current entered the tissue. The excitation does not remain localised, but is conducted to a distance, shown by the contraction of the terminal muscle. A precisely similar result is obtained with Mimosa and other sensitive plants.

I give a sketch showing these effects observed in *Biophytum sensitivum* (fig. 110); the figure to the left exhibits the excitation initiated at the kathodic point at the starting of the current, which was conducted in both directions. After the recovery of the leaflets, the circuit was broken and excitation was transferred to the point of anode.

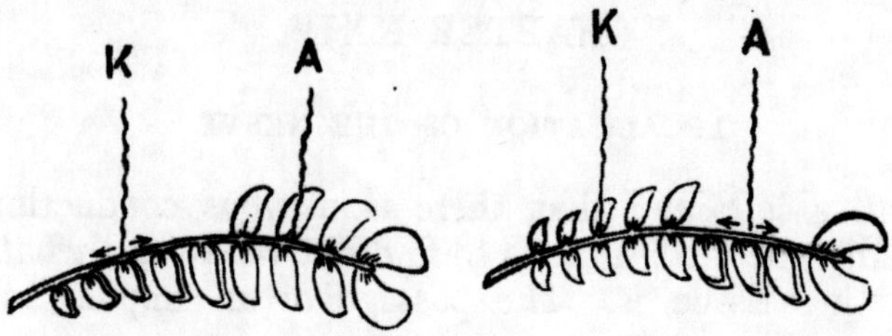

FIG. 110. Illustration to the left shows excitation induced at kathode-make; that to the right the effect of anode-break (*Biophytum sensitivum*).

Inasmuch as excitation is initiated in both animal and plant at kathode-make and at anode-break, and since the impulse in both cases is arrested by intense cold, by the application of poison, and by an electrotonic block, the inevitable conclusion is that transmission is essentially the same physiological process in both; if it be called 'nervous' in the case of the animal, there is equal reason for applying to it the same term in the case of the plant.

CHAPTER XXIV

LOCALISATION OF THE NERVE

HAVING ascertained that there is nervous conduction in the plant, the next thing is to find out where and what the conducting tissue is. The passage of an impulse along

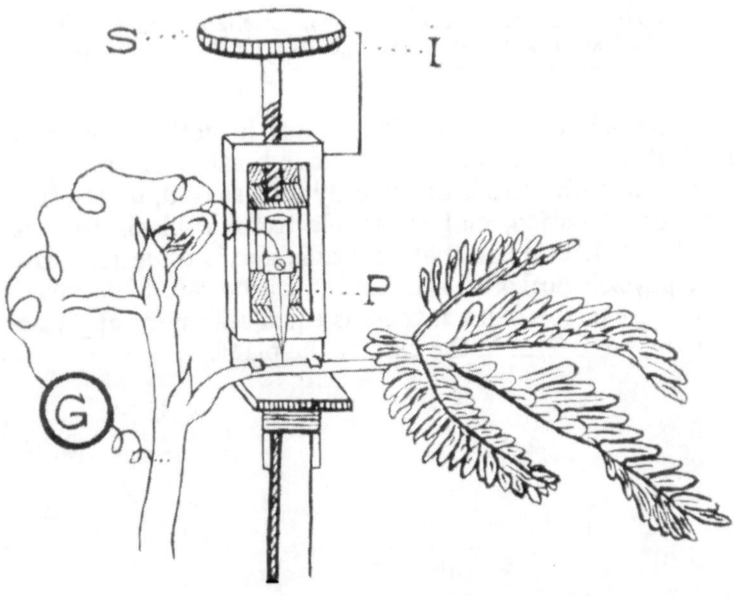

FIG. 111. The Electric Probe for localisation of Nervous Tissue in the petiole of Mimosa.

P, the probe in circuit with the galvanometer, G; S, the screw-head, by the rotation of which the probe is forced into the petiole; I, index by which the depth of intrusion may be determined.

a nerve does not produce any visible change; we can, in fact, only detect its passage by the negative electric change that accompanies it. The nerve imbedded in a non-conducting tissue may be likened to an electric cable sheathed in non-conducting gutta-percha. There may be a single conductor in the cable, or there may be two conductors. We can pick up messages going through the cable if we thrust in a metallic pin suitably connected with a galvanometer. No messages can be picked up until the pin just touches the

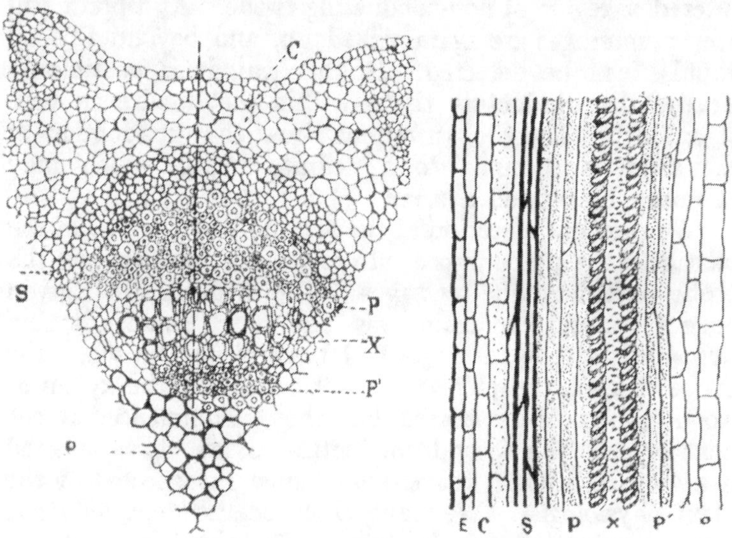

FIG. 112. Transverse and Longitudinal Sections of a Single Vascular Bundle.

Left figure: Transverse section. The dotted vertical line indicates the passage of the Electric Probe.

C, cortex; S, sclerenchyma; P, external phloem; X, xylem; P′, internal phloem; O, pith.

Figure to right: Longitudinal section of the bundle. Note elongated tubular cells, both in the external and in the internal phloem. (The section passed through one side of the bundle and not through the middle.)

conductor; the extent of intrusion of the pin tells us the depth at which the conducting strand is situated.

Acting on this principle, I have been successful in localising the nerve imbedded in non-nervous tissue (fig. 111) by making use of the Electric Probe already referred to in previous chapters. The distant end of the sub-petiole of Mimosa is periodically excited whilst the Probe is being thrust into the leaf-stalk by steps, say, of 0·05 mm. at a time. At first no electric change, indicative of nervous impulse, can be picked up; we have to probe to a greater depth. Now a message begins to be received, and at a certain depth the electric disturbance is most pronounced. We take note of this depth of intrusion. As the Probe is thrust in still deeper, the electric indication disappears; the Probe has traversed beyond the conducting tissue and entered a region of non-conducting tissue. At a point still deeper, messages are again picked up, and beyond this no impulse can be detected. In this way it is possible to localise the conducting tissue within the hundredth part of an inch. These observations show that conduction of excitation is confined to a definite tissue, which may therefore be termed a *nerve*.

We then cut a section of the leaf-stalk at the line of the passage of the probe in order to find out at what points the probe picked up the messages. The epidermis had given none; the cortex again was plainly a non-conducting wrapping; strong messages had been picked up when the Probe entered the phloem. As it passed to the xylem or wood, the messages ceased, but they recommenced at the next layer. The second conducting tissue thus detected is a second, internal phloem, until now unsuspected by the plant-physiologist. We have thus localised not only one nervous layer, but two (fig. 112). The significance of this double system of nerves, one external and the other internal, will be explained later

Nerve-distribution Mapped Out by Effective Staining

If we can discover a stain that picks out the nerve-strands, already identified as the vehicle for nervous impulse, then the nerve distribution in the plant can be clearly made out. In this way it would be possible to distinguish two neighbouring systems of tissue having different functions, or to establish the similar functions of two tissues which happen to be separated from each other. The application of hæmatoxylin and saffranin stained the nervous tissue a deep violet and made it stand out prominently from other tissues. This test confirmed the results reached by the indications of the Electric Probe; the outer and inner phloems were similarly stained, which indicates that they are, in fact, two separate nerves. In the petiole there are four such double strands, each pair starting from each sub-petiole and ending in the pulvinus.

In the stem of Mimosa itself there are two opposite main vascular bundles, each of which contains a double strand of nerves. These give off lateral branches to the leaves, thereby assuring conducting continuity between stem and leaves. An impulse initiated by stimulation of the stem can thus be sent in an outward direction to the leaves ; an impulse generated in the leaves can, on the other hand, travel inwards to the stem and be then conducted up and down to the other leaves and may even cause their fall. It will also be noted that the two main strands of conducting phloem converge and meet at the apex of the stem (fig. 113). This explains how it is that under moderately strong stimulus applied on one side of the stem, the ascending impulse crosses over at the top and becomes a descending impulse on the opposite side (see p. 184).

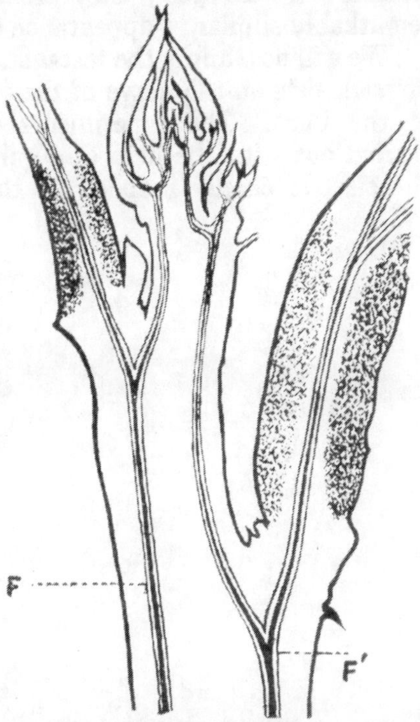

FIG. 113. Conducting Continuity between Stem and Leaves.

Longitudinal section of stem bearing leaves on each side. The two ascending bundles F, F′ give lateral branches to leaves, and meet at apex. The double phloem stained violet stands out against the background. Pulvinus of leaves shaded.

THE ISOLATED PLANT-NERVE

The phloem-strand in the vascular bundle of Mimosa is thus shown to be a nerve conducting excitation. It is impossible to pull this out from Mimosa without tearing it to pieces; I however succeeded in isolating the nerve from the leaf-stalk of a Fern. The hard casing of the leaf-stalk was broken carefully, and on pulling it apart the vascular nerve-strands were isolated; they are soft, and white in colour, remarkably similar in appearance to animal nerve (fig. 114).

We will now apply the tests generally employed by animal physiologists on the nerve of the frog, to the isolated nerve of the Fern. The experiments on the frog's nerve are carried out with the help of a galvanometer which records the electric change induced by the nervous impulse. The

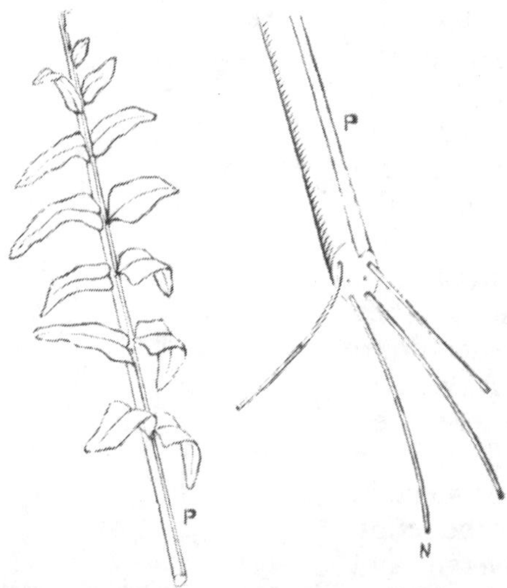

FIG. 114. Frond of Fern: the exposed vascular strands, N, are shown in the enlarged figure to the right.

electric records of the plant-nerve under varied conditions are found to be in every way similar to those of the animal nerve. The following example demonstrates this in a striking manner.

It is a well-known fact that a nerve, after lying too long idle, becomes more or less inert, and that it can be stirred into activity by being subjected to continuous stimulation or tetanisation. The feeble response of the inert nerve becomes greatly enhanced after a period of tetanisation. This is illustrated in fig. 115, in which the first three responses are those of the inert frog's nerve; after tetanisation the responses are far more pronounced than at the beginning.

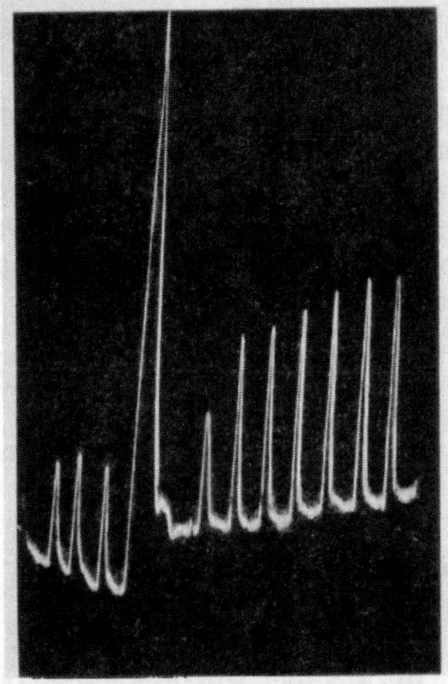

FIG. 115. Record of enhancement of amplitude of response as after-effect of thermal tetanisation, in frog's nerve.

The first three responses are normal. Brief thermal tetanisation is then applied, and the responses subsequently obtained under the original stimulation are enhanced.

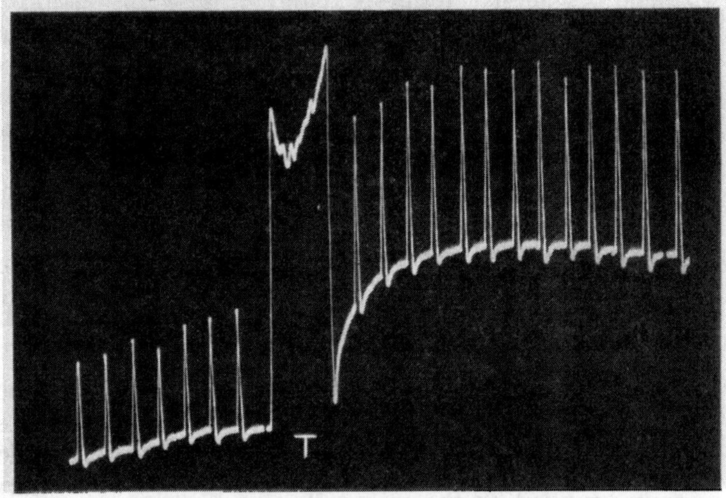

FIG. 116. Photographic record of effect of tetanisation, T, inducing enhancement of normal response in nerve of Fern.

Results exactly similar are obtained with the nerve of the Fern (fig. 116).

The Synaptic Membrane

Protoplasmic continuity was at one time regarded as essential to conduction of excitation in plants. But in the animal there is no protoplasmic continuity across the nerve-junction where neurone joins neurone, the separating membrane being known as a synapse. This synapse acts like a valve, allowing the impulse in the nerve to travel more easily in one direction than in the opposite. Certain other characteristics of the nervous impulse arise from the valve-like action of the synapse.

A detailed examination of the nervous tissue of the plant shows it to consist of elongated tubular cells, the transverse septa of which act as synapsoidal membranes.

Bahnung or Facilitation

The action of a valve will evidently be facilitated in consequence of its frequent opening by repeated stimulation. A hinge, rusty from long disuse, will require a stronger push than one that has been in frequent use which facilitates its working. A similar result is found in the transmission of nervous impulse, where 'the effect of passage of an impulse is to diminish the resistance, so that a second application of stimulus evokes the reaction more easily.'[1] This characteristic reaction is known as 'Bahnung' or opening out of the path by frequent traffic.

The characteristics of the nerve of the plant are found to be very similar to those of the animal. In the first place, though the impulse can travel in both directions, there is a preferential direction in which it travels more easily and with greater speed in consequence of the valve-action of the synapsoidal membranes. There is thus a greater facility for centrifugal than for centripetal transmission.

Again, Bahnung or facilitation is produced by previous stimulation. Thus in a typical case, the conducting power of the specimen was so low that the impulse due to the test stimulus applied on the leaf-stalk at a distance of 15 mm. from the pulvinus, failed to reach it. On application of a much stronger stimulus the impulse forced its way through

[1] Starling, *Principles of Human Physiology* (1920), p. 305.

the resistance and was transmitted to the pulvinus with considerable rapidity. The path being once made, it was easy for subsequent impulses to be transmitted, the previously ineffective stimulus now becoming effective.

Cotton-wool Treatment

The following observations as regards the growth or degeneration of nervous activity through use and disuse are suggestive. A plant carefully protected under glass from the stimulating blows of the elements looks sleek and flourishing, yet in reality it is flabby and decadent. Anatomically, the nervous tissue is present, but from want of use it is functionally inactive.

It is very interesting to watch, in a plant in this condition, the growth of nervous conduction under the influence of stimulating blows. There is at first no transmission; after a time excitatory impulses begin to be transmitted; continued stimulation enhances the conducting power to a maximum.

Here we have displayed before us the modification of the organism by its environment, the creation of the organ by the cumulative effect of stimulation. The nerve unstimulated lies passive and inert: but stimulation energises it, and its excitability and conductivity become highly exalted.

The Leaf as a Catchment-basin of Stimulus

There is a particular aspect of the action of stimulus which is of utmost importance for the maintenance of the life of the plant. For the continuance of its normal activities the internal tissues have to be maintained in an optimum tonic condition by stimulation, which can only come from outside. Among the external stimuli naturally accessible to the plant none is more potent than light. All the conditions favour the transmission of its stimulating effect to the interior along the nervous channel—the phloem in the vascular tissue. The pulsation maintaining the ascent of sap has been shown to come to a standstill when the plant is deprived of light; but after exposure to the stimulus of light the pulse-throb becomes renewed, the sap is again pumped up, and life riots through the channels which serve as arteries. The outspread leaf, in which the vascular bundles are distributed in fine ramifications, is not merely a special structure for the fixation of carbon from the carbonic acid gas in the air, but also a catchment-basin for the stimulus of light, the excitatory effect of which is gathered into larger and larger nerve-trunks for transmission to the interior of the plant. The distribution of vascular bundles in the interior is such that no mass of living tissue is too remote to be stimulated by the excitation conducted by these nervous channels.

Thus all parts of the plant are maintained by means of nerve-connection in the most intimate and rapid communication with each other. It can only be by virtue of the existence of a system of nerves that the plant constitutes a single organised whole, each part of which is affected by every influence that falls upon any other.

CHAPTER XXV

HELIOS AND HIS LEAF-CHARIOT

In the animal, rapid means of communication between its different organs is often a matter of life and death, for when it becomes aware of any threatened danger by sight or sound, an urgent message is sent along the connecting nerve to the organ of locomotion, which is immediately set in action to enable it to escape. The attitude of the creature is profoundly modified by the action of the stimulus; if this be favourable to its well-being, it turns towards it, if unfavourable, away from it.

In the plant also two opposite reactions are induced under the action of moderate and of intense stimulation. The latter endangers the life of the plant, and I shall in a later chapter describe the characteristic movements by which the Mimosa plant evades any source of intense stimulation.

I will here speak of the attitude assumed by the plant for the absorption of the stimuli which are for its well-being. It has been already pointed out that light is essential to the assimilation of carbonic acid by the plant; the greatest absorption of light take place when the upper surfaces of its leaves are perpendicular to the incident rays. This perpendicular adjustment is technically described as the dia-heliotropic attitude of the leaves.

Photographs of certain plants grown in my garden (fig. 117) show this quest for light. The middle figure is a Sunflower growing near a wall, the plant being exposed to light from the western sky. The leaves numbered 1 and 3 have undergone a twist—right-handed or left-handed—such that the upper surfaces of the leaf-blades are placed at right angles to the incident light. The figure to the extreme right shows the curvature and adjustment of a different species of Sunflower which was grown in the open. In the morning, it bent over to the east and all the leaves showed appropriate movements and torsions to face the

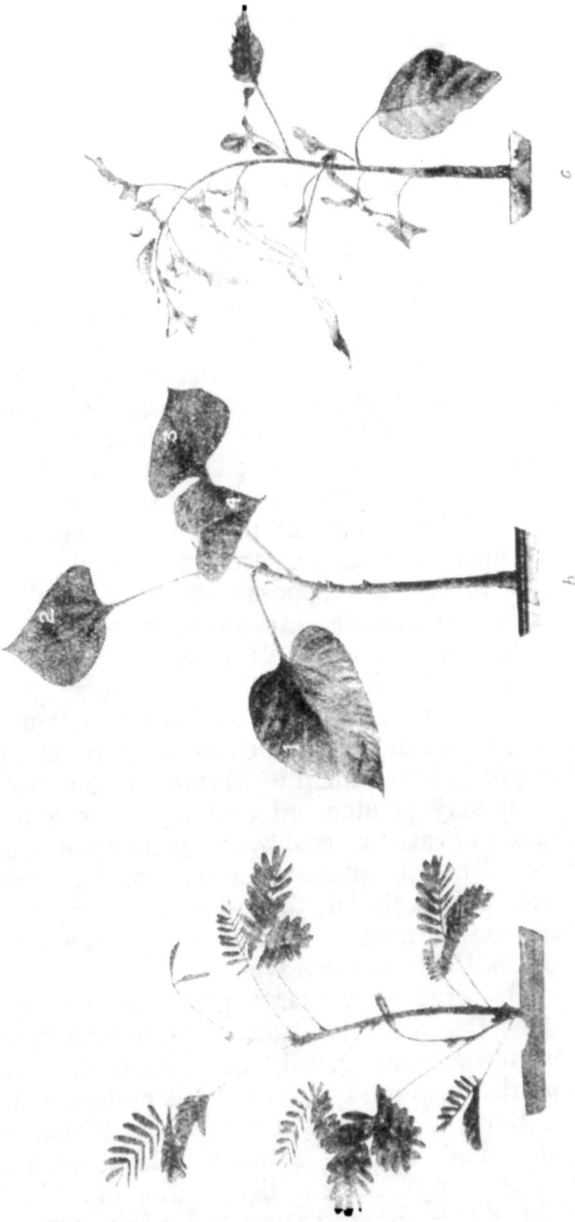

FIG. 117. Dia-heliotropic adjustment of leaves. (*a*) in *Mimosa pudica*; (*b*) in *Helianthus annuus*; and (*c*) in another species of Helianthus.

light. In the afternoon the plant bent over to the west, all the previous adjustments and torsions being completely reversed. The plant continued to exhibit these alternate swings day after day, till the movements ceased with age.

The leaf-adjustment is shown in a still more striking manner by Mimosa, a photograph of which is reproduced on the left of the illustration. The plant grown in a pot had been exposed to the northern sky. It will be seen that the leaves which directly faced the light have been raised and so placed that the sub-petioles with their leaflets are at right angles to the strongest illumination. The lateral leaves, on the other hand, have undergone appropriate twists, the plane of the leaflets being adjusted perpendicularly to the light; it will be noted that the petioles to the right and to the left have undergone opposite torsions. After the assumption of this position, the pot containing the plant was moved round 180°. This brought about a new adjustment in the course of twenty minutes, the plane of all the leaflets being once more at right angles to the light. The new adjustment necessitated a complete reversal of the former movements and torsions.

Basking in the Sun

How now are these movements produced? The leaves turn as if to bask in the sun. What happens when we turn our out-stretched palms towards the sun? In order to effect this, the very complex muscular machinery in the arm has to be set in action, producing a right-handed or left-handed twist, or a movement up or down. In the case of the leaf, some such complex movement in its muscular organ, the pulvinus, must be induced by the action of light.

For the necessary observations on the plant, I will take Mimosa, in which the motility of the pulvinus is so very pronounced. It used to be thought that the motor machinery of the leaf was simple, permitting only a single up or down movement. The results of my investigations show, however, that this is by no means the case; for the mechanism is capable of producing very complicated movements, not merely up and down, but also twists to the right and left.

The pulvinus itself may be regarded as consisting of four quadrants, left and right, upper and lower, numbered

1, 4, 3, and 2 respectively (fig. 118). Now when the left quadrant 1 is locally stimulated by a feeble electric shock or by a beam of light, the leaf answers not by a fall but by a left-handed twist. If the stimulus be transferred from the left to the right quadrant 4, the response is a right-handed torsion. Stimulation of the upper quadrant 3 produces a slow up-movement, while that of the lower quadrant 2 causes a more rapid down movement. The leaflets attached to the four sub-petioles are carried like so many flags by these muscle-like reactions of the four quadrants. If the green

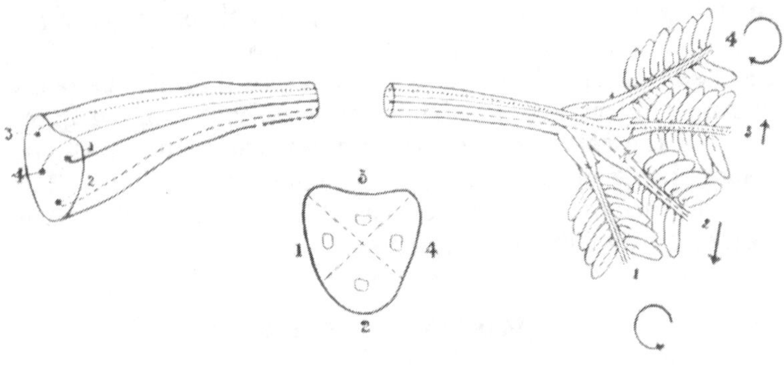

FIG. 118. Showing the course of four nerve-strands from the four sub-petioles to the pulvinus. (*Mimosa pudica.*)

The lower figure is a diagrammatic section of the pulvinus with its four quadrants. Quadrants 1 and 4, which give rise to left- and right-handed torsions, are respectively in nervous connection with sub-petioles 1 and 4. The lower effector 2 is connected with sub-petiole 2, the response being a rapid down-movement. The upper quadrant 3 is in connection with sub-petiole 3, the response being a slow up-movement. The observer is looking towards the parent stem.

leaflets carried by the four sub-petioles are kept covered so as to be sheltered from light, and if sunlight coming from the east strikes the right quadrant 4, the leaf as a whole twists to the right, carrying all the leaflets on the four sub-petioles with it, as if to face the sun which they cannot see. On the other hand, if the sunlight from the west strikes the left quadrant on the opposite side, the twist is reversed and the leaflets face the west.

So much for the turning movements caused by direct stimulation of the motile organ by light. But under natural conditions the pulvinus is shaded from direct light by the shadow cast by the leaflets. How then does the leaf turn to the light when the motile organ is not directly stimulated? Nothing simpler: by the pulvinus being made aware of the state of affairs through a message from the leaflets.

Nerve-connection between Leaf and Pulvinus

I have shown elsewhere[1] that there are definite nerve-connections between the four sub-petioles and the four quadrants. Thus a moderate electrical stimulus applied to the left sub-petiole initiates a nervous impulse which, reaching the left quadrant 1, produces its characteristic response, namely, a left-handed torsion. In the same way, stimulation of the right sub-petiole produces a right-handed torsion. Stimulation of the middle sub-petioles produces up or down movements. These results are also obtained if the sub-petioles carrying the leaflets are stimulated not by an electric shock but by the stimulus of light.

The Anchored Moth

A man rowing a boat, when pulling only one oar, circles round and round without any forward movement. It is when both the oars are pulled with equal vigour that a directive and purposeful movement can be produced. If a number of moths are anchored by means of thin threads which do not interfere with the free movement of the wings, they will be found all pointing to the light. This is because both their eyes are then equally stimulated by the light, and the nervous impulse transmitted to the wings makes them beat at the same rate and with equal vigour. Had one moth by chance been aslant to the light, then one eye would have received a stronger stimulus than the other; this would have made it turn round till both the eyes received the same amount of light. The leaf acts like an anchored moth which turns towards the light.

Thus in the plant, a slanting beam of light falling on one of the sub-petioles, say number 1 on the left, causes a twist, by which the leaf turns round till the fourth sub-

[1] 'The Dia-Heliotropic Attitude of Leaves as Determined by Transmitted Nervous Excitation,' *Proceedings of Royal Society*, B, vol. 93, 1922, p. 153.

petiole comes into the line of light: but the effect of stimulation of that sub-petiole will tend to make the leaf twist to the right, putting on a break, as it were, to the former movement. The two opposing reactions balance each other when the two sub-petioles are equally illuminated, this being the case when they are perpendicular to the incident light. The lateral adjustments of the leaf as a whole are thus made by the two sub-petioles 1 and 4, which are situated externally. The balancing adjustments up or down are made in response to nervous excitation transmitted by the two middle sub-petioles. It is clear that equilibrium is only possible when the leaf-surface, consisting of the leaflets carried by the four sub-petioles, is equally illuminated, and this can only occur when the upper surface of the leaf as a whole is perpendicular to the incident light. Thus the leaf is adjusted in space by the co-ordinated effects of four nervous impulses initiated at the perceptive region of the leaflets actuating the four motor quadrants in the pulvinus at a distance.

So we have four brothers, sun-lovers all, pledged to share whatever sunshine there may be, but each with his own pull at head-quarters for the movement of the whole. Imagine them lying with their faces to the sky. The sun ascending in the east throws a ray which touches the right-hand brother; he immediately sends the message to his quadrant of the pulvinus, which gives a twist to the right, moving not only the scout who sent the message, but the whole squad to face the sun. If the leaf tends to swing too far, scout number 4 rectifies the matter by a pull the other way.

After all, it is Helios, the Sun-god, who is the source of all movement on earth and of all living beings. It is he who pulls us out of our beds each day, who draws water from the equator to the top of the Himalayas, makes our rivers run and causes the winds to blow. In such mighty work, he does not neglect even the smallest leaf, upon which he descends and of which he makes his chariot. The four nerves of the leaf are as so many reins, by the guidance of which the chariot is raised or lowered, or made to swerve either to the right or to the left.

CHAPTER XXVI

DISCOVERY OF A REFLEX ARC IN PLANTS

When a child puts its finger on a flame, before it has had time to make up its mind to cry, the arm is automatically withdrawn. This is brought about by a wonderful reflex mechanism. The strong stimulus of the burn gives rise to an ingoing, sensory or afferent impulse, which, reaching a nerve-centre, becomes reflected and transformed into an outgoing motor or efferent impulse; this impulse travelling along a new path causes the rapid withdrawal of the hand, which is an involuntary and automatic action. During one of my lectures in London, a late-comer, finding all available space occupied, sat on what he took to be a raised seat, and immediately jumped off—it was a hot-water pipe! This was an unrehearsed experiment in reflex action!

A diagrammatic representation of the reflex arc at a nerve-centre is represented in fig. 119. S is an intense stimulus which impinges upon the surface of the skin; the sensory nerve conducts the afferent impulse A to the nerve-centre NC; the impulse now becomes reflected as an efferent impulse E and is sent along the motor nerve inducing contraction of the terminal muscle M. The two nerves, afferent and efferent, though distinct, run together and form what is called a mixed nerve.

I have been able to discover a parallel arrangement in the leaf of Mimosa. It has been shown that there are four vascular bundles, by which the four sub-petioles are put into nervous connection with the pulvinus. Investigation with the Electric Probe showed that each bundle contains two nerve-strands, one external, the other internal and therefore relatively more protected. The existence of two nerve-strands in each bundle was further proved by selective staining. What can be the object of this duplication?

In investigating this question stimulus was applied on one

of the sub-petioles, say the left, and was gradually increased step by step. The simplest and most convenient way of doing this is by increasing the intensity of an electric shock given by a coil. When the stimulus is feeble or moderate, the response, as already explained, is a left-handed torsional movement. But when the intensity of the stimulus is increased a new class of phenomena makes its appearance.

We will follow the course of the impulse generated by a moderately strong stimulus applied on the left sub-petiole. The inward passage of excitation along the sub-petiole is shown by the successive upward closure of the leaflets. The normally outspread leaflets seen from above appear vivid green; after the passage of an impulse the closed leaflets become an inconspicuous line of grey. When the impulse reaches the small pulvinus of the sub-petiole, this is made to move laterally towards its fellow, the next sub-petiole. The impulse then enters the main petiole and its passage gives no sign for the time being. After a while its arrival at the pulvinus is signalled by the fall of the leaf. The time required for the afferent or ingoing impulse to reach the pulvinus of the leaf is determined by observing the interval between the closure of the innermost pair of leaflets and the fall of the leaf.

FIG. 119. Diagram of Reflex Action in the Animal.

The work done by the ingoing impulse does not end here, for this impulse now becomes transformed in the pulvinus into an outgoing impulse which travels along a new path in the opposite direction. The outgoing impulse reaches the periphery, and its reversed direction of propagation is shown by successive closure of the leaflets of the second sub-petiole taking place in an outward direction (fig. 120). Hence it is clear there is a Reflex Arc in the leaf of Mimosa. The afferent nerve may be described as sensory, the efferent nerve as motor.

The measurement of interval of time between the fall of the leaf and the closure of the inner pair of leaflets of the second sub-petiole gives the speed of the motor impulse, which is significantly different from that of the sensory impulse. I have found that these two impulses are conducted by different nerves, the afferent or sensory impulse being conducted by the external, and the efferent or motor impulse by the internal nerve. This is the answer to the inquiry (p. 194) as to the significance of the presence of two nerves in each vascular bundle.

A

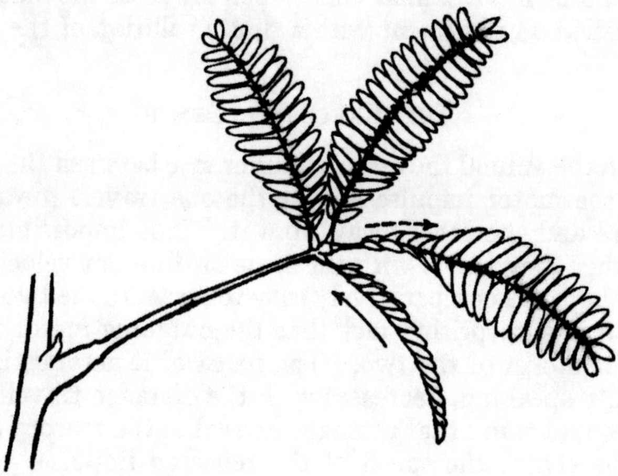

FIG. 120. **Diagram illustrating reflection of impulse.**

A, **Moderate stimulation of sub-petiole (1) gives rise to an afferent impulse which causes fall of leaf (not shown in the figure). The afferent impulse is reflected from the pulvinus as an efferent impulse which causes response of leaflets on sub-petiole (2). Afferent impulse represented by full arrow and efferent by dotted arrow.**

Under a still stronger stimulus applied on the first sub-petiole, the effect transmitted becomes more widely irradiated, and the sub-petioles exhibit response in serial succession. These characteristic features of the nervous reaction of Mimosa under increasing intensity of stimulus are remarkably similar to those of the nervous reaction in the animal, where increased intensity of stimulus enables the impulse to spread along more neurons and evoke a more widespread response.

The study of the time-relation of these reflexes shows that there is a definite interval of several seconds between the arrival of the sensory impulse at the pulvinus and the departure from it of the motor impulse. This is the 'lost time' in the reflex; it is the time occupied by the transition of the impulse from the afferent to the efferent nerve in the pulvinus, which indicates that there is a resistance or block to be overcome. Similar conditions are known to exist in the nerve-centre of the animal, where the block is entirely removed by the administration of strychnine. I find that strychnine has a similar effect upon Mimosa, the block being abolished on treatment with a dilute solution of the drug.

The Executive Centre

In the animal the essential difference between the sensory and the motor impulse is that the one travels towards the centre and the other away from it. It is impossible to tell whether they travel with the same or different velocity. In the plant it is comparatively easy to measure the two speeds, and it is a surprising fact that the outgoing motor impulse is the quicker of the two. The speed of a nervous impulse, usually speaking, decreases with the distance travelled, and one would think that after the arrival of the sensory impulse at the centre the speed of the reflected impulse would be diminished, since it has had to travel a longer distance, *i.e.* instead of travelling through the single length of the petiole it has to traverse double that length. My investigations, however, show that the speed of the outgoing motor impulse is at least six times greater than that of the incoming sensory impulse. The change of a sensory into a motor impulse in the reflex arc therefore connotes not merely a reversal in the direction of propagation, but a great discharge of energy at the centre, by which the motor impulse becomes far more intense than the feeble sensory impulse that provoked it.

It would thus appear that the centre has a specific executive function, and that a considerable amount of energy is stored there. The sensory impulse arriving at the centre pulls a trigger and the motor impulse is in consequence discharged with almost explosive intensity and rapidity. It will presently be shown that the principal function of the motor impulse is the readjustment of the outlying organs to meet a threatened crisis. There must always be a ceaseless alertness and an immediate action in response to the general need, for any disharmony means the destruction of the plant commonwealth.

Moderate stimulation, as has been explained before, makes for the well-being of the plant, such stimulation giving a healthy tone to the organism. Intense stimulation, on the other hand, is inimical to life. Let us see how a moderate or an intense stimulus affects the attitude of an animal. When a kitten is gently stroked, it expands into a round ball, purrs with delight, and is attracted towards the person who is caressing it. But when a big stick is substituted, the nature of the response undergoes a radical change. Instead of expansion there is violent contraction ; instead of joyous purring, a painful screech ; instead of attraction, a quick repulsion and straight jump for the open door. If the door be closed, the creature hides itself under the sofa. The attitude of the Mimosa likewise is characteristically modified by moderate or by excessive stimulation. The stimulus of light being of advantage to the well-being of the plant, the leaf adjusts itself so as to receive the largest amount of it. But when the light is so intense as to be harmful, a sensory message comes to the centre as a danger signal, and the order sent out to the outlying organs is for immediate withdrawal.

Lying Low

Mimosa pudica is a weed which covers large areas of ground in the tropics. The procumbent stems bear numerous leaves, so that the entire tract appears as a mass of vivid green. The danger which threatens the life of the plant is from grazing cattle, the attack of which is followed by the sudden fall of the leaves. It has been suggested that the movement of the leaves serves the purpose of scaring the cattle, though they are already accustomed to the swaying branches of trees. Moreover, the cow is not sufficiently intelligent to notice the slight movement of the leaves of Mimosa, nor would it be frightened by it. The nervous reflex may, however, subserve the protection of the plant in a different way. When one of the sub-petioles bearing leaflets is trampled on or bitten, the excitatory impulse is immediately transmitted throughout the length of the plant. In regard to the behaviour of a particular leaf, the afferent or sensory impulse causes its fall so that it presses itself close to the ground. The reflexes at the centre give rise to motor impulses which make the four sub-petioles approach each other laterally and also cause upward closure of all the leaflets. Nothing could be more startling than the rapid change, by which large patches of vivid green thus become transformed into thin lines of dull grey unnoticeable against the dull background of the soil. Like the kitten hiding under the sofa, the plant escapes danger by making itself invisible!

CHAPTER XXVII

THE CONTROL OF NERVOUS IMPULSE AND SENSATION

It is one of the greatest of all mysteries how we are brought into contact with the external world; how blows from without are felt within. Our sense-organs are like so many antennæ, radiating in various directions and picking up messages of many kinds. All of these, when analysed, are found to consist of shock-effects transmitted along different nervous channels. These shocks are caused by mechanical strokes or by impacts of aerial or æthereal waves, and we perceive them as touch, sound, or light. These messages bear, moreover, a certain potentiality to induce in us a sensation which may either be agreeable or disagreeable. The quality of the sensation is often affected by the intensity of the impinging stimulus. It is well known that while a gentle touch or a moderate stimulus of light, heat, or sound may produce a sensation which may be described as pleasant, an intense stimulus of the same nature causes a sensation which is extremely unpleasant or even painful.

The intensity of the impulse that reaches the perceptive organ depends on two factors—the strength of the external stimulus and the condition of the vehicle that conducts the impulse. Under normal conditions extremely weak stimulation gives rise to an impulse which is so feeble that it remains below the threshold of perception. Moderate stimulation gives rise to responsive sensation not unpleasant; very strong stimulation, on the other hand, causes an intense reaction of a painful character.

Our sensation is thus coloured by the intensity of the nervous excitation that reaches the central organ. We are subject to human limitations, through the imperfection of our senses on the one hand, and our over-sensibility on the other. There are happenings which elude us because the stimulus is too feeble to waken our senses; the external shock may, on the other hand, be so intense as to fill our life with pain. Since we have little power to alter the external world, is it possible to control the nervous impulse, so that it shall be exalted in one case and inhibited or obliterated in the other? Does science hold out the hope of any such possibility? This question is plainly fraught with high significance.

The Nerve-circuit

The solution of the problem of the modification of nervous impulse and the resulting sensation depends on the discovery of some means of affecting the impulse during transit, so that it can be either intensified or inhibited. How can this be effected?

The nerve-circuit may be likened to an electric circuit with transmitter and receiver connected by a conducting wire. The invisible electric impulse sent along the conductor is detected by the twitch of the needle of the receiving galvanometer. Application of extremely feeble current induces no answering twitch; the response of the galvanometer becomes perceptible and then increased under increasing intensity of the impinging electric force. When this force is very strong the answering movement of the needle becomes violent.

In the nerve-circuit parallel effects are observed as demonstrated in a more concrete way by substituting for the galvanometer needle a contractile muscle, which is also a substitute for the perceiving brain. An extremely feeble stimulus applied at the end of the nerve gives rise to an impulse which is below the threshold of response. Stimulus of moderate intensity gives rise to an impulse which causes a moderate twitch. A very strong shock gives rise to a violent contraction of the muscle; the magnitude of the response measures the intensity of the transmitted nervous impulse.

Problem of Nervous Control

There is much resemblance between the conduction of electric impulse by a metallic wire and that of excitatory impulse by a nerve. In the metal the power of conduction is constant, and the intensity of the electric impulse simply depends on the intensity of the electric force that is applied. If the conducting power of the nerve were constant, then the intensity of the nervous impulse and the resulting sensation would depend entirely on the intensity of the impinging stimulus. In that case, modification of sensation would be an impossibility. But there may be a possibility that the power of conduction possessed by a nerve is not constant but is capable of change, so that resistance to the passage of an impulse can either be decreased or increased. Should this surmise prove correct, then we would arrive at the momentous conclusion that sensation itself is modifiable, whatever be the external stimulus.

The modification of nervous impulse can be attempted in either of two ways. In the one case, we may render the nervous path super-conducting so that the impulse due to a subliminal stimulus may be brought into sensory prominence. In the other case, we may block out the pain-causing impulse of intense shock by rendering the nerve a non-conductor.

Under a narcotic the nerve becomes a non-conductor and we may thus save ourselves from pain. But such heroic measures are only resorted to in extreme cases, as when we are under the surgeon's knife. In actual life, we are confronted with unpleasantness without previous notice. A telephone subscriber has the evident advantage that he can switch off the connection when the message begins to be unpleasant. Statesmen and politicians have been known to cultivate convenient deafness; but that is mere pretence, for unpleasant remarks continue to rankle. Few have the courage of Mr. Herbert Spencer, who openly resorted to his ear-plugs when his visitor became tedious.

Molecular Transmission of Impulse

The thrill produced by stimulation of the receptive outer end of the nerve is sent inward from point to point of the conducting thread as a molecular disturbance. Imagine a row of billiard balls touching one another. A blow given to the first ball is communicated to the next and next, till the last ball starts off from the rest in acknowledgment of the blow. None of the other balls appears to have been moved from its place. In the same way each particle of the nerve, though keeping its own place, passes on the impulse.

It is evident that conduction will depend on the mobility of the particles along which the impulse is sent. Imagine two successive transmissions in the same nerve under similar stimulations; if the mobility of the particles have been increased under certain treatment, the impulse will be quickened and become more intense. Contrariwise, with diminished mobility, conduction by the particles will be weakened, even to extinction. The mobility of the particles and the power of conduction are increased or decreased by use and disuse. The nerve unstimulated lies passive and inert; but subjected to stimuli, it becomes energised and highly conducting. Stimulation by thought increases our power of thought.

The intensity of the nervous impulse causing sensation is thus not merely conditioned by the strength of the stimulus, but modified by the variable condition of the nervous particles.

Effect of Molecular Predisposition on Nervous Impulse

A further step in the reasoning carries us from mere propagation of molecular disturbance constituting the impulse, to the acceleration or inhibition of the impulse under opposite molecular predispositions. Instead of a row of balls, let us imagine a row of standing books. A certain intensity of blow applied, say, to the book on the extreme right, causes it to fall to the left, hitting its neighbour and making the other books topple over in succession. If the books had previously been tilted slightly to the left, a disposition would have been given to them which would bring about an upset under a feebler blow and accelerate the speed of the impulse. A tilt in the opposite direction would, on the other hand, be a predisposition to retard or inhibit the impulse.

Is it at all possible to induce opposite molecular dispositions in the nerve, so that in one case a sub-liminal stimulus will be effectively transmitted and brought into sensory prominence, and, in the other case, an intense wave due to a violent stimulus will be inhibited during transit and thus obliterated? These theoretical anticipations must now be subjected to experimental tests.

The molecules of the nerve have to be coerced at one's bidding into arranging themselves favourably or unfavourably for the passage of impulse. This can be effected by the employment of an electric force of a polar character. The polarising action of an electric current is shown by the fact that during its passage through a trough of water from left to right, the hydrogen atoms turn towards the right; when the current is reversed, they turn to the left. The molecules of the nerve can be similarly orientated during the passage of an electric current, so as to predispose them to facilitate or oppose the nervous impulse.

The results obtained are of extraordinary interest. By conferring on the plant-nerve a favourable molecular disposition, a feeble stimulus previously below the threshold of perception now produces an extraordinarily large response. Conversely, an intense excitation generated by a violent stimulus is arrested during transit by inducing opposite molecular disposition on the nervous tissue. The plant now takes no account of the intense shock.

A climax was reached when by the application of similar methods I was able to confer on the animal nerve a supra-conducting or non-conducting property at will. Thus under a particular molecular disposition of the nerve, the experimental frog responded to a stimulus which had hitherto been below its threshold of perception. Under the opposite disposition, the violent spasm under intense irritation caused by application of salt, NaCl, was at once quelled as if by magic.[1]

[1] 'The Influence of Homodromous and Heterodromous Current on Transmission of Excitation in Plant and Animal,' *Proceedings of Royal Society*, B, vol. 88, 1914.

Power of Directive Control by the Will

The foregoing experiments have demonstrated that two opposite molecular dispositions can be induced in nerve by the polar action of a constant electric current, accelerating or inhibiting the passage of a nervous impulse. The question naturally arises whether or not the action of the Will upon the nerves of the body may not be of a similar nature. Now, full scientific attention has not yet been given to the power of our will in controlling all bodily functions. The muscular system, for example, has been classified as voluntary or involuntary, the latter being regarded as beyond the control of the Will. Yet I have known instances in which involuntary muscles have been brought under the control of human will, so much so that the automatic beats of the heart were brought to a stop and afterwards renewed. I have also seen X-ray photographs demonstrating normal and reversed peristalsis of the alimentary canal produced at will. Very few have realised how great becomes the power of the Will when intensified by practice and concentration. There can be no doubt of the predispositions which can be conferred on the nerves by the internal power of the Will facilitating or inhibiting nervous impulses. The effect of attention or expectation in enhancing perception is familiar, so also is the power of suggestion. The following incident will show the influence of Will in controlling the nerves.

I had gone on an expedition to the borders of the Himalayan Terrai in Kumaun. Terror prevailed among the villagers, for a man-eating tigress had come down from the forest and numerous had been the toll of lives exacted. The murderess of the forest would come out even in broad daylight and leisurely take her prey from among them. When all hope of deliverance had nearly vanished, the villagers appealed to Kaloo Singh, a simple peasant who possessed an old matchlock. With this primitive weapon and with the entreaties of his fellow villagers ringing in his ears, Kaloo Singh started on his perilous adventure.

The tigress had killed a buffalo and left it in the field. Kaloo Singh waited there for her return to the kill; there was not a tree near, only a low bush behind which he lay crouched. After hours of waiting, as the sun was going down, he was startled by the sudden apparition of the tigress, within six feet of him. Trying to raise his gun, he could take no aim, as his arm was shaking with uncontrollable fear.

Kaloo Singh explained to me afterwards how he succeeded in shaking off his mortal terror: 'I quietly said to myself, "Kaloo Singh, Kaloo Singh, who sent you here? Did not the villagers put their trust in you?" I could then no longer lie in hiding and I stood up, and then something strange happened. All trembling went, and I became as hard as steel. The tigress, with eyes blazing and lashing tail, crouched for the spring; only six feet lay between us. She sprang and my gun went off at the same time: she missed her aim and fell dead by my side.'

Man Victorious over Circumstances

In the determination of sensation, then, the internal stimulus of the Will may play as important a part as the shock from outside. And thus, through the inner control of the nerve, the character of the resulting sensation may become profoundly modified. The external, then, is not so overwhelmingly dominant, and man is no longer a passive agent in the hand of destiny. He has a latent power which will raise him above the terrors of his inimical surroundings. It rests with him whether the channels through which the outside world reaches him shall at his command be widened or closed. It should thus be possible for him to catch those indistinct messages that have hitherto passed by him unperceived; or he may withdraw within himself, so that in his inner realms the jarring notes and the din of the world no longer affect him.

From the plant to the animal, then, we follow the long stairway of the ascent of Life. In the high spiritual triumph of the martyr, the ecstasy of the saint, we see the higher and higher expression of that evolutionary process by which Life rises above and beyond all the circumstances of the environment, and fortifies itself to control them.

The thrill in matter, the throb of life, the pulse of growth, the impulse coursing through the nerve and the resulting sensations, how diverse are these and yet so unified! How strange it is that the tremor of excitation in nervous matter should not merely be transmitted but transmuted and reflected, like an image on a mirror, into a different plane of life in sensation and in affection, in thought and in emotion. Of these, which is the more real, the material body or the image which is independent of it? Which of these is undecaying, and which beyond the reach of death?

Many a nation has risen in the past and won the empire of the world. A few buried fragments are all that remain as memorials of the great dynasties that wielded the temporal power. There is, however, another element which finds its incarnation in matter, yet transcends its transmutation and apparent destruction: that is the burning flame born of thought which has been handed down through fleeting generations.

Not in matter, but in thought, not in possessions nor even in attainments, but in ideals, is to be found the seed of immortality.

APPENDIX

THE BOSE INSTITUTE

THE SURGE OF LIFE[1]

In pursuit of my investigations, I was unconsciously led into the border region of physics and physiology, and was amazed to find boundary lines vanishing and points of contact emerging between the realms of the Living and Non-Living. Inorganic matter was found to be anything but inert; it also was athrill under the action of multitudinous forces that played on it. A common reaction seemed to bring together metal, plant, and animal under a general law. They all exhibited the phenomena of fatigue and depression, together with possibilities of recovery and of exaltation, yet also that of permanent irresponsiveness which is associated with death.

The further and fuller investigations of the many and ever-opening problems of the nascent science which includes both Life and Non-Life are among the main purposes of the Institute I am opening to-day. The present time may be regarded as inopportune for any new undertaking, when a great tragedy is hanging over human destiny. But it is at such a crisis that men learn to discriminate the Real from the Unreal, so that they can dedicate themselves to the pursuit of Truth which is eternal.

THE TWO IDEALS

There are at this moment two complementary and not antagonistic ideals before the country. There is first the individualistic ideal of winning success in all affairs, of securing material efficiency and of satisfying personal ambition. These are necessary, but by themselves they cannot ensure the life of a nation. Such material activities have brought as their fruit great accessions of power and wealth. There has been a feverish rush even in the realm of science for exploiting

[1] From the Inaugural Address on the occasion of the dedication of the Bose Institute, November 30, 1917.

THE BOSE INSTITUTE

applications of knowledge, not so often for saving as for causing destruction. In the absence of some power of restraint, civilisation is now trembling in an unstable poise on the brink of ruin. Some complementary ideal there must be to save man from that aimless rush which must end in disaster. He has followed the lure and excitement of some insatiable ambition, never pausing for a moment to think of the ultimate object for which success was to serve as a temporary incentive. He forgot that far more potent than competition is mutual help and co-operation in the scheme of life.

Nevertheless there have been some who, beyond the immediate and absorbing prize of the hour, have sought for the realisation of the highest ideal of life, not through passive renunciation, but through active struggle. The weakling who has refused the conflict, having acquired nothing, has nothing to renounce. He alone who has striven and won can enrich the world by giving away the fruits of his victorious experience.

The ideal of giving, of enriching, in fine, of self-renunciation in response to the highest call of humanity is the other and complementary ideal. The motive power for this is to be found not in personal ambition but in the effacement of all littleness, and in the uprooting of that ignorance which regards anything as gain which is to be purchased at another's loss. This I know, that no vision of truth can come except in the absence of all sources of distraction, and when the mind has reached the point of rest. And for my disciples I call on those very few who will devote their whole life with strengthened character and determined purpose to take part in that infinite struggle to win knowledge for its own sake and see Truth face to face.

Advancement and Diffusion of Knowledge

The work already carried out in my Laboratory on the response of matter, and the unexpected revelations in plant-life, foreshadowing the wonders of the highest animal life, have opened out very extended regions of inquiry in Physics, in Physiology, in Medicine, in Agriculture, and even in Psychology. These inquiries are obviously more extensive than those customary either among physicists or physiologists, for they demand interests and aptitudes hitherto more or less divided between them. In the study of Nature there is a necessity for a dual view-point, an alternating yet rhythmically unified interaction of biological thought with physical studies, and of physical thought with biological studies. The future worker with his freshened grasp of physics, his fuller conception of the inorganic world as indeed thrilling with 'the promise and potency of life,' will redouble his former energies for work and thought. He will thus be in a position to winnow the old knowledge with finer sieves, to re-search it with new enthusiasm and subtler instruments. His handling of these will at once be more vital and more kinetic, more comprehensive and more unified.

The advancement of science is the principal object of this Institute, as well as the diffusion of knowledge. We are here in the largest of all the many chambers of this House of Knowledge, its Lecture Room. In adding this large auditorium, I have sought permanently to associate the advancement of knowledge with the widest possible civic and public diffusion of it; and this without any academic limitations, henceforth to all races and languages, both men and women alike, and for all time coming. The lectures given here will announce, to an audience of some fifteen hundred people, the new discoveries, which will be demonstrated for the first time before the public. We shall thus maintain continuously the highest aim of a great seat of learning by taking an active part in both the advancement and the diffusion of knowledge.

It is my further wish that, as far as the limited accommodation would permit, the facilities of this Institute should be available to workers from all countries. In this I am attempting to carry out the traditions of my country, which, so far back as twenty-five centuries ago, welcomed all scholars, from different parts of the world, within the precincts of its ancient seats of learning, at Nalanda and at Taxilla.

The Outlook

With this widened outlook we shall not only maintain the highest traditions of the past, but also serve the world in nobler ways. We shall be at one with it in feeling the common surgings of life, the common love for the good, the true, and the beautiful. In this Institute, this Study and Garden of Life, the claim of Art has not been forgotten, for the artist has been working with us, from foundation to pinnacle, and from floor to ceiling of this very hall. And beyond that arch, the Laboratory merges imperceptibly into the Garden, which is the true laboratory for the study of Life. There the creepers, the plants, and the trees are played upon by their natural environments—sunlight and wind, and the chill at midnight under the vault of starry space. There are other surroundings also, where they will be subjected to chromatic action of different lights, to invisible rays, to electrically charged atmosphere. Everywhere they will transcribe in their own script the history of their experience. From his lofty point of observation, sheltered by the trees, the student will watch this panorama of life. Isolated from all distractions, he will learn to attune himself to Nature; the obscuring veil will be lifted, and he will gradually come to see how throughout the great ocean of life community outweighs apparent dissimilarity. Out of discord he will realise the great harmony.

These are the dreams that have woven a network round my wakeful life for many years past. The outlook is endless, for the goal is at infinity. The full realisation cannot be through one life or one fortune, but through the co-operation of many lives and many fortunes. The possibility of a fuller expansion will depend on very large endowments. But a beginning must be made, and this is the genesis of the foundation of this Institute. I came with nothing and shall return as I came; if something be accomplished in the interval, that will indeed have been a privilege. What I have, I will offer; and one who had shared with me the struggles and hardships that have had to be faced has wished to bequeath all that is hers for the same object.

Wider Synthesis

The excessive specialisation in modern science has led to the danger of losing sight of the fundamental fact that there can be but one truth, one science, which includes all the branches of knowledge. How chaotic appear the happenings in Nature! Is Nature a Cosmos in which the human mind is some day to realise the uniform march of sequence, order, and law? India through her habit of mind is peculiarly fitted to realise the idea of unity, and to see in the phenomenal world an orderly universe. It was this trend of thought that led me unconsciously to the frontiers of different sciences and shaped the course of my work in its constant alternation between the theoretical and the practical, from the investigation of the inorganic world to that of organised life and its multifarious activities of growth, of movement, and even of sensation. Thus the lines of physics, of physiology, and of psychology converge and meet. And here will assemble those who would seek oneness amidst the manifold.

INDEX